Data Analytics
in Reservoir Engineering

Society of Petroleum Engineers

Richardson, Texas, USA

Disclaimer

This book was prepared by members of the Society of Petroleum Engineers and their well-qualified colleagues from material published in the recognized technical literature and from their own individual experience and expertise. While the material presented is believed to be based on sound technical knowledge, neither the Society of Petroleum Engineers nor any of the authors or editors herein provide a warranty either expressed or implied in its application. Correspondingly, the discussion of materials, methods, or techniques that may be covered by letters patents implies no freedom to use such materials, methods, or techniques without permission through appropriate licensing. Nothing described within this book should be construed to lessen the need to apply sound engineering judgment nor to carefully apply accepted engineering practices in the design, implementation, or application of the techniques described herein.

ISBN: 978-1-61399-820-5 [Print]
ISBN: 978-1-61399-821-2 [Mobi (Amazon)]
ISBN: 978-1-61399-822-9 [Epub (iTunes)]
ISBN: 978-1-61399-823-6 [WebPDF (ADE)]

10 9 8 7 6 5 4 3 2 1

Society of Petroleum Engineers
222 Palisades Creek Drive
Richardson, TX 75080-2040 USA

http://store.spe.org
service@spe.org
1.972.952.9393

Table of Contents

Preface

"Reservoir engineering is difficult. The most successful practitioner is usually the engineer who, through extensive efforts to understand the reservoir, manages to acquire a few more facts and thus needs fewer assumptions."—P. L. Essley Jr. (1965)

Data analytics is a process of collecting, cleansing, transforming, and modeling data to discover useful information that can be used to make recommendations for future decisions.

Data analytics is fundamentally transforming several industries such as retail marketing, telecom, insurance, and banking. In this digital age, it is becoming more important for companies to leverage technology to change the way they operate, aided by data analytics. In the recent years, there also has been a growing application of digital technologies in oil and gas exploration and production. Oil and gas operations are becoming increasingly more complicated with modern facility infrastructures, complex reservoirs, increased regulatory requirements, changing workforce demographics, the fast pace of unconventional-field development, and the competitive landscape.

In this book, we focus on the impact of data analytics on reservoir engineering applications—more specifically, the ability to characterize reservoir parameters, analyze and model reservoir behavior, and forecast performance to transform the decision-making process.

Why Is Data Analytics Relevant Now for the Oil and Gas Industry? The confluence of several factors such as sensor explosion, advances in cloud and hardware technology, and innovations in data science algorithms, in addition to the recent downturn in the oil and gas industry alongside the success of data analytics in other industries, has contributed to the crossroads where we are with respect to applying data analytics in reservoir engineering work processes.

Over the past few years, several successful case studies (Mehta 2016; Sankaran et al. 2017) have demonstrated the benefits of applying data analytics to transform the traditional reservoir model to a data-driven decision support. The key questions that remain are related to determining the right work processes that lend themselves to data-driven insights, how to redesign them effectively in the new paradigm, and adopting the appropriate business model to complement them. Most oil and gas companies have already embarked on this journey and are at varying maturity levels on this trajectory.

How Does Data Analytics Add Value in Reservoir Engineering Applications? The rapid progress of big data and data analytics offers companies opportunities to automate high-cost, complex, and error-prone tasks. Many oil and gas operators are progressively accelerating efforts to capture these opportunities in order to reduce costs and increase efficiency and safety. Companies that adequately employ automation can significantly improve their bottom line by converting data into information and enabling timely decision making.

Data analytics applications in reservoir engineering can add value to various types of reservoirs; in particular, a data deluge due to the scale and pace of field development has characterized the rise of unconventionals.

While physics-based methods such as numerical simulations and analytical modeling remain in use, they present major challenges for unconventional assets, in particular

- Lack of reliable conceptual models to properly describe the underlying physics.
- Difficult characterization of the inputs required.
- Complex physics-based models require long run times, which conflicts with the short decision cycles in most unconventional developments.

The computational requirements of a physics-based model often lead to a trade-off between accuracy and model footprint. Today, it is still impractical to develop and maintain a basinwide simulation model that is accurate at the well level. It is essential however to understand the key factors driving the economic performance of unconventional-field development. This gap is often addressed in practice by data-driven models designed to support field development decisions such as optimal well spacing, targeting, and completions design.

Operators and software companies have extended the utility of data-driven models to support transactional decisions regarding entering or exiting unconventional plays. Emerging plays and appraisal stages are characterized by significant uncertainty regarding economic viability, where data-driven models in conjunction with systematic field pilots through experimental design are used to derive early insights and reduce risk. Subsequently, data-driven models are used for field development decisions and to optimize drilling and completions practices.

Data collection programs to assess rock and fluid properties (e.g., fluid, log, core acquisition) have also benefited through application of data analytics. Instead of collecting extensive fluid or core samples and conducting laboratory experiments, empirical correlations and data-driven methods are used to extract key features and estimate fluid or rock properties.

One of the most important tasks of a reservoir engineer is to make production forecasts. When the governing equations describing the underlying subsurface behavior are reasonably well-understood such as in conventional reservoirs, data analytics is used to accelerate production forecasting through proxy models (response-surface models), reduced-physics models (e.g., capacitance/resistance models), or reduced-order models (e.g., proper orthogonal decomposition, trajectory piecewise linearization). More recently, data-driven and physics-constrained predictive uncertainty analysis methods have been developed to accelerate reservoir management decisions through a direct forecasting approach without ever building an actual model. In cases where the underlying phenomenon is not well-understood or is very complex such as in unconventional applications, data-driven methods (i.e., regression methods in machine learning) are used to map the inputs directly to desired response variables (such as cumulative production or estimated ultimate recovery at the end of a time interval). Often physics-inspired features or other features are used to improve production forecast accuracy, such as the new variants of decline curve analysis (Holanda et al. 2017).

A key responsibility of the reservoir engineer is reservoir management, which includes responsibly developing and producing fields. Different stages of field development require different objectives and specific analyses. During early field life, the focus is on assimilating key reservoir data, (e.g., pressure, saturation, fluid distribution, rock and fluid properties, and hydraulic connectivity) and understanding subsurface behavior (e.g., reservoir connectivity, drive mechanisms, tendency for sand production). In this stage, data analytics can be used to augment data collection programs and accelerate continuous learning by monitoring reservoir response through automation. The vast amounts of data available from the sensors enable modern artificial intelligence methods to derive insights. Realizing production optimization goals in real time requires fit-for-purpose models that can collect available data, analyze, and act at the relevant time frequency. This is becoming practical with data-driven methods for problems such as detecting well productivity decline and identifying contributing factors (Sankaran et al. 2017), recommending well workover candidates for production enhancement opportunities, and estimating injection rates and well controls to optimize recovery.

Data-driven analytics is also used in enhanced oil recovery (EOR) and improved oil recovery (IOR) applications as a screening tool to accelerate lengthy evaluations. In waterflood and steamflood applications, several successful case studies have demonstrated the efficacy of using data-driven and hybrid models to maximize production on a daily basis, select shutoff candidates, and subsequently optimize overall field development.

How Can We Overcome the Challenges to Successfully Apply Data Analytics? Successful data analytic applications in reservoir engineering begin with a fundamental understanding of the business needs and of the key physics at play. Availability of adequate data that are characterized by good quality is essential for building a robust model and solution. However, data quality remains a big challenge for several companies to produce sustainable solutions, resulting from a variety of data management issues that need to be addressed. Outliers and missing, duplicate, obsolete, and unstructured data are just a few of the challenges that must be overcome. Additionally, multiple sources of disparate data need to be integrated into a single consistent version with contextual information. Several companies have embarked on this effort by constructing data lakes and establishing data standards and good data management practices to enable this transformation. Fundamentally, organizations are realizing the value potential of data and need to pay attention to what data are collected and how data are collected, processed, and stored for both current and future applications.

In routine operations, data coverage often is limited to narrow operating ranges. For instance, wells are drilled in reservoirs with favorable rock properties, unconventional wells are completed with relatively less variation of completions design, and so on. This limits the ability to use these data to develop robust data models that can be adequately extrapolated. It requires careful attention to plan how data are collected through experimental design methods, keeping in mind the type of analysis that needs to be performed.

In general, there is a perceived sense of lack of know-how for approaching data analytic projects in reservoir engineering applications. This is exacerbated by a shortage of adequate staff that merges data science and engineering skills. Having a good grounding in data science and in the underlying physical processes will help to assess the validity of the analytics approach and to interpret results appropriately. Further, these new methods modify existing work processes and will require appropriate change management for user adoption and realizing the benefits. Having management support is essential for driving these deep-rooted changes leading to business transformation.

Several of these challenges are not unique to the oil and gas industry. Other industries such as retail, banking, insurance, and healthcare have successfully leveraged big data to drive efficiency and growth profitably (Hand 2007; Raghupathi and Raghupathi 2014; Sarwar et al. 2017; Goldstein 2018). The oil and gas industry is looking at learning from others and extending this success to oilfield operations. Beyond reservoir engineering, several other technical areas (such as drilling, completions, and geoscience) that are characterized by data processing and interpretations could significantly benefit from data analytic methods.

The authors would like to thank those that peer-reviewed our book and provided us with feedback prior to publication, Eduardo Gildin, Hector Klie, Shahab Mohaghegh and Suryansh Purwar.

About the Authors

Sathish Sankaran is EVP of Engineering and Technology at Xecta Digital Labs. Prior to that he served as Engineering Manager of Advanced Analytics and Emerging Technology for Anadarko Petroleum Corporation. His work focuses on modeling and optimizing hydrocarbon production from reservoir to process plant, with emphasis on blending physics and data-driven methods.

Sebastien Matringe is currently Director of Reservoir Engineering at Hess Corporation. He previously held various leadership and engineering positions at Newfield Exploration, Quantum Reservoir Impact and Chevron. He holds a "Diplome d'ingenieur" in Fluid Mechanics from ENSEEIHT in France and MS and PhD degrees in Petroleum Engineering from Stanford University.

Mohamed Sidahmed serves as Machine Learning and Artificial Intelligence R&D Manager for Shell. He also serves as Director on the Board of Petroleum Data-Driven (PD^{2}A) Technical Section of the SPE and Program Evaluator for ABET, dedicated STEM PEV contributing to the profession. There he contributes to the quality of technical education in collaboration with professionals from academia, industry and government.

Xian-Huan Wen is a Chevron Fellow and the Chapter Manager of Reservoir Simulation Research and Optimization in Chevron Technology Center. Wen holds a PhD degree in Water Resources Engineering from the Royal Institute of Technology, Sweden and a PhD degree in Civil Engineering from the Technical University of Valencia, Spain. He has authored or coauthored more than 80 papers and holds two US patents.

Luigi Saputelli, is a Reservoir and Production Engineering Expert with 30 years of experience. He held various positions around world including PDVSA, Hess, Halliburton and ADNOC since 2014. He has issued over 100 publications comprising integrated reservoir and production management to digital automation projects, co-authored three books and nine patents. Saputelli is an active SPE volunteer, underwriting over 60 SPE events and technical committees. He is also the founder of Frontender, a petroleum engineering firm based in Houston, TX.

Andrei Popa is currently Reservoir Management Consultant within Chevron's Upstream Capability Reservoir Management unit. In addition to working for Chevron, Popa is an Adjunct Associate Professor at University of Southern California for the last 11 years, where he teaches Advanced Natural Gas Engineering. He has more than 23 years of experience leading Artificial Intelligence and Machine Learning projects and cross-functional teams focused on delivering complex optimization solutions and conceptual models across Chevron enterprise. He has served on the SPEI Board of Directors (2015 – 2018) as WNA Regional Director. He has published more than 80 papers on AI technologies and co-authored the book *Artificial Intelligence & Data Mining (AI&DM) Applications in the Exploration and*

Production Industry. He is the recipient of 2013 SPE International Management and Information Award, and 2015 SPE International Distinguished Service Award. Popa earned his PhD and MS degree in Petroleum and Natural Gas Engineering from the West Virginia University.

Serkan Dursun is Leader of Artificial Intelligence Center of Excellence CoP in Hydrocarbon Management Division at Saudi Aramco. He was Principal Data Scientist at Marathon Oil, Data Scientist at Schlumberger and Senior Technologist at Halliburton. Dursun served as Adjunct Professor at University of Houston. He has 11 publications in IEEE & SPE and holds nine US patents. He has special interest in Artificial General Intelligence (AGI), Conversational AI, and Explainable AI (XAI).

DATA ANALYTICS IN RESERVOIR ENGINEERING

Sathish Sankaran, Editor and Co-Chair
Sebastien Matringe, Editor and Co-Chair
Mohamed Sidahmed, Luigi Saputelli,
Xian-Huan Wen, Andrei Popa,
Serkan Dursun

Peer Reviewers:

Eduardo Gildin, Hector Klie, Shahab
Mohaghegh, Suryansh Purwar

1. Introduction

Data analytics has been broadly applied to a variety of areas within the oil and gas industry including applications in geoscience, drilling, completions, reservoir, production, facility, and operations. However, we limit our discussion here primarily to reservoir engineering applications.

1.1. Objectives. The key objectives of this book are as follows:

- Describe the relevance of data analytics for the oil and gas industry, with particular emphasis on reservoir engineering.
- Outline the methodology and guidelines for building robust models for reporting, diagnosis, prediction, and recommendations.
- Provide an overview with examples of successful applications of data analytics in reservoir engineering and illustrate advantages and pitfalls these methods.

1.2. Organization of this Book. Section 2 outlines the main objectives of this book and introduces data analytics. Successful applications in other industries are provided as a reference to illustrate the digital transformation that is driven by data and other new technologies.

Section 3 provides an introduction to the data driven modeling methods and the various stages in the modeling life cycle.

Section 4 illustrates the nontechnical elements needed to assimilate the results of data driven methods into decision making for businesses. The emphasis here is on business processes and human factors required to enable the successful adoption of data analytics methods.

Section 5 discusses a number of published practical applications of data analytics in reservoir engineering, spanning fluid analysis, core analysis, production surveillance, reserves and production forecasting, reservoir surveillance and management, EOR and IOR, reservoir simulation, and unconventionals. We discuss several reservoir modeling approaches as a spectrum of possibilities between full-physics and data-driven methods that incorporate data analytics to various degrees.

Section 6 discusses future trends that address needed advancements in areas related to data, applications, and people for successful adoption of data analytics methods in the long term.

1.3. Background. While several other industries have experienced significant benefits, wide adoption of data-driven analytics is still in its fledgling stages in the oil and gas industry. Typical reported benefits include reducing costs, improving decision quality, gaining business agility, and driving operational efficiency improvements. McKinzey (2016) estimates key value drivers from applying digital technologies contributing 10–40% reduction in maintenance costs, 10–20% cost of quality reduction, 30–50% reduction in downtime, and 45–55% productivity increase through automation of knowledge work.

Fig. 1.1 shows digital maturity across industries (Grebe et al. 2018) and, in particular, shows that the energy industry is in the nascent stages of digital acceleration. Although early, the promises and potential of digitization and analytics to reservoir engineering applications are very exciting.

In a recent industry report (from GE/Accenture), surveys show that 81% of senior executives believe that big data analytics is one of the top three corporate priorities for

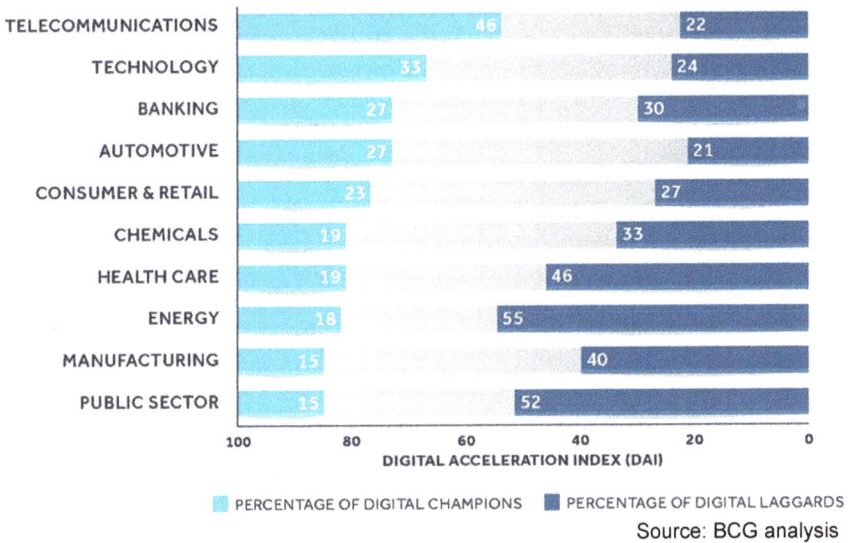

Fig. 1.1—Digital maturity across industries (Grebe et al. 2018).

the oil and gas industry (Mehta 2016). There seems to be a rapid increase in uptake and sense of urgency by several companies to implement data analytic solutions. The major driving factors are the need to improve well performance to reduce costs and major breakthroughs in digital technology that enable commercial viability. As a result, the digital-oilfield market is projected to reach USD 28 billion by 2023 (Market Research Engine 2018), to which data analytics is a major contributor.

Companies have formed new data science organizations and recruited a new breed of technical professionals to solve complex oil and gas problems. This trend is primarily driven by current market conditions that drive companies to become more efficient, in the footsteps of other industries.

The confluence of the following factors has led to the crossroads where we are with respect to applying data analytics in reservoir engineering work processes:

- Development and wide availability of inexpensive sensors that have accelerated subsurface and surface data collection (i.e., velocity, variety, and volume of data)
- Advancements in modern data storage and management (including cloud infrastructure)
- Breakthroughs in hardware technology that address massively parallel computations [with central processing units (CPUs) and graphical processing units (GPUs)]
- Innovations in data science algorithms that leverage modern hardware and availability of large volumes of data to improve accuracy
- Recent industry downturn and success of unconventional reservoirs (shale and tight oil/gas reservoirs, called "unconventionals") in the oil and gas industry that have renewed focus on operational efficiency
- Proven success of data analytics methods to transform other industries

Business operations are becoming increasingly digital in many industries. Other industries (e.g., banking, healthcare, insurance, power utilities) are not simply creating a digital strategy, they are digitizing their business strategy.

Banking. The finance industry has been an early adopter of big data and data analytics to drive revenue and reduce risk in the areas of high-frequency trading, pre-trade decision-support analytics, sentiment measurement, robo-adviser, antimoney-laundering, know your customer, credit risk reporting, and fraud mitigation (Hand 2007; Srivastava and Gopalkrishnan 2015, Cerchiello and Giudici 2016).

Healthcare. Healthcare analytics describes actionable insights that have been undertaken because of analysis of data collected from claims, pharmaceutical R&D, clinical data, and patient behavior for reducing rising costs and providing better benefits across the board in the areas of evidence-based medicine, predictive computational phenotyping, drug screening, and patient similarity (Raghupathi and Raghupathi 2014; Archenaa and Mary Anita 2015; Sarwar et al. 2017; Muhammad et al. 2017; North American CRO Council 2017).

Insurance. With data analytics, insurance businesses have increased accessibility to huge volumes of data that can be converted into customer insights resulting in improved profitability and overall performance in the areas of personalized customer service, fraud prediction, and accident likelihood (Wuthrich and Buser 2017; North American CRO Council 2017).

Power Utilities. Aging infrastructure and demand challenges have forced the power utility industry to leverage data analytics to reduce cost and improve reliability, with (for example) smart meters, smart grid technology, and power-outage predictions (Guille and Zech 2016; Goldstein 2018).

Some common goals among these industries embracing big data and data analytics involve enhanced customer experience, cost reduction, improving operational efficiency, and value optimization. Data-driven insights are increasingly driving decision making across these businesses. Other typical benefits reported include cross-functional agility, cost resilience, speed, and innovation.

However, there are significant challenges for adapting these technologies by the oil and gas industry. It begins with a proper understanding of what is data analytics and what it can do for the enterprise user.

1.4. What Is Data Analytics? Data analytics is a process of collecting, cleansing, transforming, and modeling data to discover useful information that can be used to make recommendations for future decisions.

There are four major maturity levels of data analytics.

Descriptive Analytics. Describe the problem on the basis of past and current data to understand what is happening. Real-time dashboards or reports are commonly used to mine the analytics.

Example. Automatically identify productivity decline in wells and present with supporting evidence.

Diagnostic Analytics. Analyze the past and current data to investigate the cause of a problem and explain why it happened. Results are often expressed as diagnostic charts or visuals.

Example. Diagnose root cause of failures by correlating operating parameters with equipment downtime.

Predictive Analytics. Analyze likely scenarios to determine what might happen. Predictive forecasts are generated for likely outcomes.

Example. Compute production forecast ranges for a new completion design in a different geological area for an unconventional well.

Prescriptive Analytics. Reveals what actions should be taken to maximize utility. This is the pinnacle of analysis that makes recommendations for optimal decisions.

Example. Automatically compute recommendations for field operations on optimal operational policy for waterflooding (i.e., adjusting water-injection rates).

Considerable amounts of data are being generated during exploration, development, and operation of oil and gas reservoirs. The amount of data produced by oil and gas fields by various disciplines now exceeds our ability to effectively treat the data. The volume, variety, and velocity of data that are being generated pose a tremendous challenge to effectively store, process, analyze, and visualize the data in reasonable timespans.

Data analytics is often used interchangeably with other terms such as data mining, statistical learning, machine learning, or artificial intelligence. This is often an exploratory approach in contrast to physical models. However, there are several structured processes and proven methods that can be used to develop robust models for analysis.

1.5. What Is New in Data Analytics? While statistics is an essential part of the science of data analytics, the latter goes far beyond that to include concepts and practices from artificial intelligence and machine learning among others. In the past decade, a transformation has occurred to shift to new systems for processing data and the ways data get studied and analyzed. In 2013, IBM estimated that 90% of the data in the world had been created in the preceding 2 years.

Advances in open-source, nonrelational databases and cloud technologies enabled overcoming the "scalability" challenge that was plaguing the statistical methods. This coincided with the ability to manipulate and process vast data sets with powerful hardware and provided a breath of fresh air to a certain class of machine learning algorithms that are inherently data hungry.

As a result, the number of data science projects exploded, spawning more innovation in new algorithms and clever tweaks to existing algorithms. This gave rise to useful discoveries of data-driven modeling methods that could do a better job of analyzing data and detecting nonlinear relationships and interaction between elements.

Today, a functional data scientist defines the problem, identifies key sources of information, designs frameworks for collecting and screening needed data, explores data for relationships, evaluates various modeling strategies for analytics, productionizes the data models, and establishes systems to monitor the health of the model.

1.6. What Value Can Data Analytics Create for the Oil and Gas Industry? Companies now realize that data constitute a vital commodity and the value of data can be realized through the power of data analytics (Saputelli 2016). Leveraging hidden insights from mining data can help the oil and gas industry make faster and better decisions that can reduce operational costs, improve efficiency, and increase production and reservoir recovery. Data analytics can thus play an important role in reducing the risks inherent in the development of subsurface resources. These analytic advantages can improve production gains by 6 to 8% (Bertocco and Padmanabhan 2014).

While data analytics has broad applications in reservoir engineering, the vast number of wells and pace of operations in unconventionals allow data to play a critical role in the decisions that create value.

Field data-collection programs (such as fluid, logs, core) are augmented with data-driven models to interpolate across the entire field. Not only does this reduce operating costs, but this could also leverage a fit-for-purpose method where physical models are complex or cumbersome (Rollins and Herrin 2015).

Automated mapping programs built using big data solutions enable companies to calculate unconventional oil and gas reserves across vast geological areas and a large number of wells in a fraction of the time it takes traditional manual methods (Jacobs 2016). These tools also allow companies to identify refracturing candidates and compare completion designs with offset operators across lease lines.

Continuous data-driven learning allows new wells to be brought online with better performance and reduced cycle time by optimizing drilling, targeting, well spacing, and completions parameters.

In conventional reservoirs, modern data-driven methods provide a pragmatic method to handle vast amounts of sensor data that provide an early-warning system for subsurface issues (Wilson 2015; Sankaran et al. 2017). This form of predictive analytics can help companies assess their portfolio efficiently for better business decisions, such as acquisitions, divestitures, and field development.

1.7. What Are the Challenges? While there is a large explosion in the application of data analytics to several reservoir engineering problems, there is still a lack of common understanding and established guidelines for building robust and sustainable solutions.

Business understanding is fundamental to a successful data analytics project. Proper background in the underlying physical (and nonphysical) phenomena helps assess the situation appropriately. It helps in validating assumptions, constraints, risks, and contingencies appropriately. This also helps to determine proper data analytics project goals.

Data Issues. Data-driven discovery often requires validation and interpretation based on fundamental understanding of the data sources, data quality, and the analysis process (Hayashi 2013). Data quality remains a key challenge for a number of companies, stemming from ad hoc data management practices, lack of data standards, and multiple versions of truth.

Subsurface data are fundamentally laden with uncertainty owing to sparse samples (e.g., fluid, logs, core) collected in the field. While temporal data from sensors (e.g., in a well) might be ubiquitous, the data might be geospatially sparse (e.g., in a reservoir). In some cases, the data format might not be conducive for real-time data analytics. In addition, proper data selection for modeling and analysis is often poorly understood.

Standards to format or store the data are often proprietary or nonexistent, leading to challenges in integration and quality control. Data quality is often in question, forcing each engineer to personally investigate data sets and check the data quality, a task that is often repeated multiple times in the same company over the years.

Data integration and repurposing data with contextual information is often not an easy task. Several companies have now embarked on creating in-house data foundations to address these issues and enable big data analytics.

Model Selection. One of the big challenges is in knowing which modeling techniques would work best for a given problem. Sometimes, this is addressed through an exploratory data analysis and trying a variety of methods through trial and error.

Some modeling methods naturally lend themselves to easy understanding, whereas others are confounded and not easily interpretable. The applicability of certain algorithms might also depend on the amount of data available or the amount that is required to be processed.

Data selection and modeling assumptions are fundamental to build robust models that can be used for decision making. Model evaluation and business success criteria should be clearly defined to have successful implementations. Further, it is not sufficient to answer the question once, but create a solution to empower practitioners to use data in new ways.

Skills Shortage. The technical skills that are required to combine data science, engineering, geoscience, and solve exploration and production business problems are in short supply (Rassenfoss 2016). Additionally, work products created by these data analytics programs often need redesigned work processes and cultural shift for realizing the digital transformation.

In the oil and gas industry today, the practice of data analytics is relatively new and, as such, job responsibilities are somewhat vague with respect to the expectations. Often, there is a lack of consistency in the expectations and definition of the roles between a data scientist, data engineer, business analyst, and a programmer. While some prefer people with specific industry experience, others look for talent outside the industry.

Whereas, several of these challenges are not unique to oil and gas industry, other industries have leveraged big data to drive efficiency and growth. Business leaders in the oil and gas industry are looking toward learning from others and extending this success to oilfield operations.

Summary

- *Other industries have reported success in adopting data analytics, with benefits such as reducing costs, improving decision quality, gaining business agility, and driving operational efficiency improvements.*
- *Advances in hardware technologies, cloud computing, data management, and new algorithms and the explosion of data have earmarked the new age of data analytics over the past few years.*
- *Several recent applications of data analytics in reservoir engineering have emerged, with potential to reduce operational costs, improve efficiency, and increase production and reservoir recovery.*
- *There are key challenges to the successful application of data analytics in reservoir engineering and its adoption—namely, data issues, choice of modeling methods, shortage of skills, and proper balance between physical understanding and data-driven methods, among others.*

2. Data-Driven Modeling Methodology

A variety of terminologies have been used in the industry to describe data analytics (sometimes loosely), such as data science, statistical learning, machine learning, and artificial intelligence (AI).

Data analytics and data-driven models in general refer to a collection of tools encompassing data collection, validation, aggregation, processing, and analysis for extracting insights from the past, predicting future performance, and recommending actions for optimal decisions on the basis of possible outcomes. Techniques used in data-driven models can include computational intelligence, statistics, pattern recognition, business intelligence, data mining, machine learning, and AI (Solomatine and Ostfeld 2008).

The process of discovering insightful, interesting, and novel patterns—as well as descriptive, understandable, and predictive models—from large-scale data is at the core of data-driven models (Zaki and Wagner 2014). The main goal is to extract important patterns to gain insights from historical (or training) data using supervised or unsupervised learning methods. In supervised learning, a functional relationship is "learned" between a parameter of interest and several dependent variables on the basis of training data or representative cases. The learned model can then be used to predict the outcome given a different set of inputs. In unsupervised learning, associations (or patterns between inputs) are learned using techniques such as cluster analysis, multidimensional scaling, principal-component analysis, independent component analysis, and self- organizing maps. These methods provide a useful way for classifying or organizing data, as well as understanding the variation and grouping structure of a set of unlabeled data.

2.1. Modeling Strategies. A model is a means to describe or understand a system that can sometimes be used to predict future outcomes. **Fig. 2.1** shows a model of a mathematical description of the static or dynamic behavior of a system or a process, where a set of equations is used to transform known input parameters to predict outcomes of output parameters. The process of identifying the best model that defines the relationship between the inputs and outputs is known as "learning" (also known as history matching, model calibration, or training in the reservoir engineering literature).

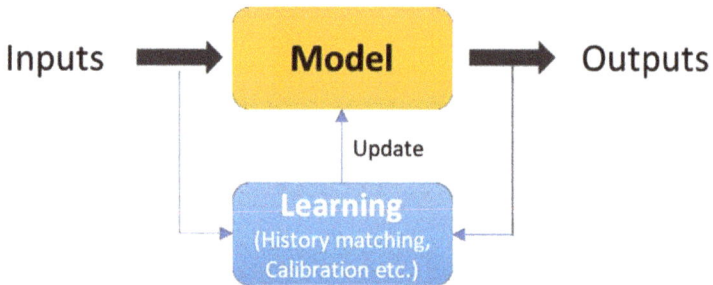

Fig. 2.1—Model overview.

In general, modeling approaches in reservoir engineering can be broadly classified into the following types:

- Analogs and scaled models
- Full-physics models
- Reduced-physics models
- Reduced-complexity models

- Data-driven models
- Hybrid models

Analogs and Scaled Models. A scaled model is generally a physical representation of the reservoir (rock and fluid) that maintains accurate relationships between important aspects of the model, although absolute values of the original properties need not be preserved. This enables demonstrating the physical phenomenon reasonably in miniature. Typical examples of these methods include fluid laboratory experiments in pressure/ volume/temperature (PVT) cells to determine fluid properties and coreflood experiments to determine rock properties or rock/fluid interaction parameters.

Analogs are a method of representing information about a source system (target reservoir or field) by another particular system (source reservoir or field). In cases where necessary information is not available to sufficiently model the reservoir mathematically, analogs can be used to predict outcomes such as reservoir recovery or production profiles.

Full-Physics Models. These are first-principles-based models that require a fundamental understanding of the underlying phenomena with the processes involved so that they can be represented mathematically in terms of physical equations and numerically as reservoir simulators. While it is impossible to model all the detailed underlying physical mechanisms, these are often referred to as full physics models and are solved using numerical methods. As a part of this process, several physical parameters (often high dimensional) are needed to adequately characterize the system, which are obtained through laboratory or field tests, through empirical correlations, or through calibration with field observations. When the underlying processes are complex, formulating these physics-based models is often cumbersome, laborious, expensive, or infeasible with limited resources for practical purposes. For example, fluid flow through multistage hydraulically fractured horizontal wells in unconventional reservoirs encompasses modeling of fluid flow in a network of rock matrix and fractures (natural and induced), with coupled multiphysics processes including geomechanical effects, water blocking, stress-dependent rock properties (permeability, porosity), Darcy or non-Darcy flow, adsorption/ desorption, and multiphase effects.

Reduced-Physics Models. When the full-physics models are cumbersome to build (and calibrate) or not fast enough for the intended modeling objective, or when all the multiphysics are not well-understood, a reduced-physics model can be built. Typically, this involves simplification of the physical process through some assumptions or modeling only portions of the physics such as material-balance models, streamline simulation, neglecting pressure-dependent variability of properties, and INSIM and capacitance/resistance models (CRMs) (Sayarpour 2008; Chen et al. 2013; Cao et al. 2014; Cao et al. 2015; Holanda et al. 2015; Holanda et al. 2018; Guo and Reynolds 2019). If the dominant physics is captured, these models (CRMs) can still be used under a variety of conditions with reasonable accuracy and are much faster. Reduced-physics methods are well-suited for production forecasting and reservoir surveillance methods that require frequent computations that are based on the latest information.

Reduced-Complexity Models. While full-physics models are more explanatory, they are often computationally intensive and high dimensional (i.e., they have several degrees of freedom that might not be well-constrained through adequate field observations). A class of techniques that reduces the model complexity can be used to create a low-dimensional model that is faster to compute and still preserves the major dynamics of the modeled system (e.g., upscaling, reduced-order modeling). Typically, these models are derived from the full-physics model—i.e., the high-dimensional model is still the starting point in the generation of the reduced-complexity models (He 2013; He et al. 2011a, 2011b, 2013, 2015a, 2015b, 2016a, 2016b). However, the model order reduction is associated with an accuracy loss that needs to be balanced against the computational acceleration and the ultimate model objectives (Chen et al. 2013).

Data-Driven Models. These are often built using data alone and require an understanding of dependent and independent variables. Advancement of machine learning techniques such as representational learning opened new frontiers and enabled AI systems to rapidly adapt to new tasks, with minimal human intervention (Goodfellow et al. 2016). A precise understanding of the physical processes at play is not required. Insights into the processes are derived directly from the data by analyzing patterns. Correlation is not identical to causality, but it can often lead to better understanding of causality. Recognizing recurring patterns can target an area for investigation and help understand the cause-and-effect process. However, data-driven analytics are often limited to patterns seen in the historical training data and perform poorly when extrapolated to new operating states. Under such circumstances, the models can be updated easily through retraining over new patterns. The speed of data-driven models under favorable conditions is often very attractive, and if they tip the advantage in favor of a sound and timely asset decision, they can add substantial business value (Gentil 2005; Al-Anazi and Gates 2010; Wicker et al. 2016; Courtier et al. 2016). Fluid and core analysis methods are rooted in empirical data, and data-driven methods lend themselves naturally to develop better models for these applications. Additional examples for production forecasting, reservoir management, and EOR are discussed in Section 5.

Hybrid Models. Taking the best characteristics of physics-based and data-driven modeling approaches, hybrid models use physics-inspired relationships between inputs and outputs but allow for flexibility and use low-dimensional models to capture localized effects. Sometimes, the starting point of hybrid models is detailed first-principles models, which are then reduced to simpler forms through simplifications, feature extraction, or field observations. Note that simplification often leads to a less general model but can be used under special considerations. There is an increasing interest in systematically deriving and using hybrid models that are more robust in handling new (unseen) patterns and still offer substantial speed advantage for online applications (Klie 2015). Some of the early research in reinforcement learning showed that some hybrid approaches could require only an approximate model, and only a small number of real-life cases. The objective is to successively "ground" the data-driven policy evaluations using physical evidence, but to rely on the approximate model to suggest local change (Abbeel et al. 2006). Hybrid models might also be particularly useful in multiscale, multiphysics systems.

There are two new approaches that are gaining popularity to develop hybrid models:

- Physics-inspired feature engineering—Transform inputs in a physically meaningful manner to be used as features for data-driven modeling methods.

- Physical consistency-based regularization—Incorporate physical principles as consistency (or inconsistency) metrics as a regularization term in the objective function as part of the data-driven modeling methods.

2.2. Model Development. Data-driven models are typically developed iteratively with progressing levels of complexity. Initial implementations might focus on a skeleton version of the proposed model or only span a part of the system. Later iterations include extensions to accommodate a more complete or a more detailed model. Typically, the data-driven model development process involves the following steps.

1. Define the objectives and purpose for the model.
2. Analyze the system to be modeled to understand the scale, complexity, and processes involved.
3. Collect, validate, analyze, and prepare the data set available for modeling.
4. Define acceptable accuracy of the model.
5. Select model and proposed modeling approach.
6. Develop a learning model and a modeling workflow.
7. Calibrate model against held out data.
8. Validate model against blind data.
9. Deploy model for intended purpose.

Repeat Steps 6 through 8 until desired accuracy is achieved, or a satisfactory termination criterion is met.

Repeat Steps 7 through 9 if the model is used to support operations (i.e., keep the model continuously refreshed for use over long periods of time to maintain high accuracy).

It is imperative to revisit the above steps if the business or operating requirements change. See Appendix A for a detailed description of the model development process.

2.3. Enabling Technologies. Generally, upstream and subsurface processes are complex. The presence of such levels of complexity makes it sometimes advisable to build a single global model that adequately captures the system/process behavior. A workaround is segmentation of relatively rich training data into several subsets and building separate specialized models on each segment. The models serve as local or expert models, and this type of modular model development gives rise to more-accurate representation of complex systems in the form of a committee machine (Sidahmed and Bailey 2016).

To ensure continuous improvement, new and enhanced modeling algorithms should be considered. Use of emerging techniques and selection of the best approach should be piloted and assessed against similar competing models by developing multiple competing models.

High-performance and in-memory analytics was shown to shorten data-driven-modeling cycle time and enabled smart workflows on big data (Holdaway 2014). The emergence of new frameworks such as deep structural and hierarchical learning transformed the way complex physical systems such as the subsurface are modeled. Deep learning (DL) is regarded as a class of machine learning techniques that exploit hierarchical layers of nonlinear transformations for supervised or unsupervised feature extraction, pattern analysis classification, and prediction. One characterization of this paradigm of models is the multiplicity of stages of nonlinear transformation and learning of feature representation at successively higher, more-abstract layers (Deng and Dong 2014).

Key enablers for the advancement and success of data-driven modeling techniques include the following primary factors:

- Acceleration of computer processing abilities such as GPUs
- Richness and rapid expansion of data-set size used for training
- Recent advances in machine learning and DL algorithms

2.4. Uncertainty Quantification and Mitigation. Uncertainty is a well-recognized problem in modeling. It is multifaceted, being derived from observation and measurement errors, system variability, ambiguous or conflicting knowledge, lack of complete understanding about the system's behavior now and in the future, and the modeler's beliefs and experience. It is important that these uncertainties be addressed during model design and development and be conveyed in association with the results. The workflow within which the model will be used should be designed to quantify the sensitivity of the results to the uncertain inputs so that decisions can take appropriate account of the sensitivity.

Trained data-driven and machine learning models are used to make predictions for new data drawn from a relatively similar underlying distribution. The process of using the model is distinct from the process that creates the model. All models are bounded, to various degrees, to the environment they were initially designed to operate within. This implies that any attempt to use a model designed within specified operating conditions for a significantly different or changed environment is likely to fail.

Several confidence measures are put in place to ensure that model results are valid and acceptable within an agreed boundary. The upper and lower boundaries of a model outcome provide an adequate measure for quantifying risk associated with output. Confidence assessment measures used during model development include

- Classical confidence intervals
- Bayesian confidence intervals
- Asymptotic intervals
- Internal confidence intervals

To control for the risks that could affect confidence in model output, it is imperative to establish an adequate governance process for effective model risk management.

Summary

- *Reservoir modeling methods fall on a spectrum of modeling strategies ranging from full-physics to data-driven methods, including reduced-physics, reduced-complexity, and hybrid models.*
- *The modeling approach is determined primarily by its purpose, data availability, speed, accuracy, and the interpretability requirement.*

- *The life cycle of continuous model development and deployment should be considered and can be enabled by supporting technologies (e.g., high-performance computing, in-memory analytics).*

3. Decision Making with Data-Driven Models

While several companies are investing in analytics, very few have been able to generate the desired kind of business impact so far. To benefit from the full data analytics value chain, it must be aligned with the overall decision-making process.

Upstream companies make routine capital allocation decisions (e.g., development, operational, mergers and acquisitions) across a diverse portfolio of assets.

- Development decisions—e.g., number of wells (or well spacing), well locations, sequence of well drilling, completions practices, secondary or tertiary recovery
- Operational decisions—e.g., well stimulation, recompletion, choke management, chemicals program
- Commercial decisions—e.g., acquisition or divestiture of portfolio assets

Understanding the performance of an existing asset or the potential payback of a prospect is an essential part of this evaluation (Harding 1996; Denney 2002; Fernandez et al. 2012; Grønnevet et al. 2015; Kaushik et al. 2017). Typically, every company periodically works up these forecasts and economic analyses as part of their planning operations, which is often tedious and difficult. However, the value of these analyses can only be realized if they result in effective decisions that are properly executed (Jafarizadeh and Bratvold 2009).

Fig. 3.1 shows the sequence of the typical decision-making process.

- Collect—Establish systems to access all relevant data, observations, prognosis, and prior information.
- Analyze—Automated and/or manual processes to assess all available information.
- Forecast—Predict future performance (potential) under expected constraints and operating conditions.
- Deliver—Deliver and implement decision within the asset.
- Monitor—Evaluate validity of forecast against actual performance.
- Verify—Perform retrospective studies to improve decision-making process.

Fig. 3.1—Decision-making process.

Note that the decision-making process is often cyclical. As new data become available, they expand the range and scale of analysis. For example, when an exploration well is drilled, it provides a large quantity of information (e.g., logs, fluids, pressure) to analyze and understand the nature of the reservoir. Further, when a subsequent appraisal well is drilled, it provides different perspectives on reservoir continuity, fluid gradients in the reservoir, delineation limits, and other factors.

Competitive advantage can be gained by reaching a better decision through focused intelligence, a faster decision by reducing cycle time, or a less-expensive decision by lowering associated costs. It is also essential to benchmark asset performance against (internal and external) competitors to refine strategic planning and growth initiatives. Having this micro- and macrolevel intelligence using fit-for-purpose reservoir and well performance models forms the cornerstone of successful organizations making sound decisions.

For this purpose, predictive analytics technologies allow for more fact-based decision making (Mohaghegh et al. 2017; Villarroel et al. 2017; Wilson 2017), rather than relying solely on human intuition. These analytic advantages could help oil and gas companies improve production from aging fields and sustain production plateaus from newer assets. Companies with better analytics capabilities are more likely to be in the top quartile for financial performance, make decisions faster than their peers, and execute decisions as planned (Brulé 2015 and David 2016).

However, a common challenge is to sift through the abundance of data and find ways to filter out superfluous information to identify major problems or key drivers. Some recent efforts are being made in the communications protocol to enable real-time decision making such as Energistics Transfer Protocol (ETP). ETP is a new communications protocol specification that enables efficient real-time data transfer between rigsite, office, and applications. It facilitates data exchange through data standards such as WITSMLTM (for drilling, completions, and interventions data), PRODMLTM (for production data from reservoir/wellbore boundary to custody transfer point), and RESQMLTM (for reservoir characterization, and earth and reservoir models). However, successful implementation of these standards requires cooperation and discipline between vendor and operating companies.

It is also difficult sometimes to define the path to value creation and the implications of technology strategy, operating model, and organization. Often, oilfield data are dirty, uncertain, and of poor quality. Therefore, data analytics might not be suited for all types of problems and needs to be addressed carefully.

As the development and application of data analytics in decision-making processes increase, there is also a growing need for proper validation and review processes to ensure quality of the delivered products. Further, the audit trail and traceability of the decision to the underlying data are important. While there are more-well-established procedures in most operating companies to understand, evaluate, and determine the usefulness of traditional (physics-based) modeling approaches, data analytic models are seen as black boxes. Therefore, it is important to establish model-validation processes that are clearly defined and well-understood by all stakeholders during the deployment of these models. Promoting the interpretability and transparency of these models is also an important consideration.

An effective approach to address this problem for decision making is presented in the following sections.

3.1. Value Creation. The centerpiece for creating value starts with identifying the most valuable applications, which lend themselves naturally to data analytics.

In unconventionals, the vast number of wells, insufficient understanding of underlying complex physics, and rapid pace of operations allow data to play a critical role in the decisions that create value. Analytic capabilities allow operators to collect and analyze subsurface data for the following:

- Basinwide characterization to improve geological interpretation that supports actual well performance
- New-well delivery through continuous learning and optimizing the recipe for drilling parameters, well spacing, and completions techniques as new wells come online to improve overall field performance

In conventional reservoirs, data analysis allows quick understanding of reservoir behavior such as reservoir connectivity, performance degradation, and identifying drive mechanism and estimating ultimate recovery. Aiding this process is the downhole instrumentation and sensors in modern assets that allow real-time surveillance of reservoir performance and equipment health. Early identification of well productivity issues and reservoir performance can then be addressed through appropriate mitigation steps that reduce downtime and improve overall recovery.

3.2. Organizational Model. Knowing the value inherent in better analytics helps focus the efforts. This must be complemented by appropriate organizational structure to make effective decisions. A good organizational structure allows decisions to be made at the right time through cross-functional collaboration and puts the right data in the hands of decision makers.

Considering the unconventional example again, this requires several key functions such as land, regulatory, subsurface, drilling, completions, and operations to work together in a fast-paced environment. Each function might generate and analyze a large amount of data, but an operating model is needed to weave together the data for a shared view. A common understanding of data (single version of truth) and clear visibility into activities are required to have an integrated view of what is happening in the field.

Some companies address this challenge by deploying an asset-based organizational model, rather than a functional one. In this model, all key functions are deployed into one (often, geographically based) organizational structure. In other cases, with larger companies, some common functions are organized as a centralized team that serves the assets. Regardless of the model that a company adopts, it needs to create a pathway for collaboration among functions, with better systems and processes that allow not only for rapid and integrated sharing of data and insights from analytics, but also for organized decision making with clear lines of accountability.

Decision making is often influenced by the organizational culture—manifested by the sum of all values recognized and practiced in the company. A data-driven culture empowers all levels of employees to use it in their respective functions. Organizations often need to overcome human factors and align critical data key performance indicators to appropriate business processes.

3.3. Execution. Another key ingredient for effective decision making using analytics is to develop the right capabilities and talent to make the most of the data. Hiring and retaining strong analytic talent is challenging for the oil and gas industry. This is often a scarce resource, and the talent profile in demand is not typically found within oil and gas information-technology (IT) functions.

Further, the data foundation that supports high throughput volumes and modern analytics engines requires new IT skills to manage cloud and open-source architectures. Companies with strong data analytics capabilities balance domain knowledge (i.e., geoscience, engineering) with analytic skills and are focused on solving specific problems and identifying new opportunities.

Finally, change management of stakeholders is an essential but often overlooked component in this value chain. Changes in the decision-making process must be properly adopted by personnel at all levels for it to transform the business.

Building a well-oiled execution machinery, where decisions are moving faster, aided through appropriate data-driven and traditional analysis, takes time and investment. Therefore, this requires sustained focus by top management.

Summary

- *The decision-making process (e.g., development, operational, commercial) must be completely aligned with the data analytics value chain to realize the full benefits.*
- *Data analytics might not be suited for all types of problems, especially when the data are limited or of poor quality, and if the results are not easily interpretable or transparent for decision making.*
- *An effective approach to ensure data analytics is useful for decision making is to selectively identify key value-adding business processes supported by the right organizational structure and focus on execution strategy to achieve desired outcomes.*

4. Reservoir Engineering Applications

This section discusses several applications of data analytics that focus on characterizing reservoir parameters, analyzing reservoir behavior, and forecasting performance.

4.1. Fluid Pressure/Volume/Temperature. Fluid PVT analysis is the study of the phase behavior of oilfield hydrocarbon systems. Phase behavior refers to the behavior of vapor (natural gas), liquid (oil, water), and solids (asphaltenes, wax, hydrates) as a function of pressure, temperature, and composition. The terms PVT and phase behavior are used interchangeably.

Characterization of PVT properties is fundamental to understanding fluid flow through reservoir rocks and flow pipes, including calculation of oil and gas reserves, production forecasts, estimating efficiency of EOR methods, surface separator design, and flow assurance.

Establishing the phase behavior of petroleum fluids is usually expensive and challenging. Fluid samples have to be taken carefully and preserved adequately; a meticulous series of experiments and analyses have to be performed to establish the phase behavior of these fluids under varying pressure and temperature conditions (McCain 1990; Danesh 1998; Whitson and Brulé 2000; Pedersen et al. 2014).

In contrast, some simple fluid properties, such as density and gas/oil ratio (GOR), can readily be determined at minimal expense. Standing (1947) published the first set of PVT correlations based on crude samples collected from California reservoirs. His seminal idea was to use readily available fluid properties such as GOR, oil and gas gravity, pressure, and temperature to estimate more-complex properties, such as the saturation pressure or formation volume factors. Various authors concerned with gas, bubblepoint, and dewpoint fluids have extended this seminal work over the following decades to cover several basins worldwide.

Initial attempts focused on discovering empirical correlations manually or intuitively that best described the desired fluid properties either as nonlinear parametric regressions or through slide-rule charts (see Lasater 1958; Nemeth and Kennedy 1967; Standing 1979; Vazquez and Beggs 1980; Glaso 1980; Hanafy et al. 1997; Dindoruk and Christman 2001; Okoduwa and Ikiensikimama 2010; Moradi et al. 2010; Bandyopadhyay 2011; Valle Tamayo et al. 2017).

A common element across the above references was that the independent variables remained mostly the same. The authors primarily attempted to find better correlations that fit different fluid sample data sets across various hydrocarbon basins. While some authors recognized the errors introduced through the various processing steps including field sample collection, laboratory measurements, and contamination, there are not many details available on systematic methods used to address these errors as part of their analyses. **Fig. 4.1** shows measurement error for PVT properties such as GOR.

Fluid Property	Uncertainty in PVT Measurement (%)
GOR	3–10
B_o	1–4
μ_o	10–30
Condensate/gas ratio	5–15
B_g	1–4
μ_g	10–30

Fig. 4.1—Typical measurement error for PVT properties (Alboudwarej and Sheffield 2016).

Fig. 4.2 shows where Valkó and McCain (2003) analyzed worldwide oil samples (1,743 sample records) and published absolute average relative error (AARE ≈ 12 to 45%) in bubblepoint-pressure estimation by several popular empirical correlations. In addition, some correlations worked better for certain fluid properties and fluid types than for others. This led to data-driven estimation of PVT properties to extract key features and reduce prediction errors.

As data-driven modeling techniques evolved, several authors have proposed the use of advanced machine learning techniques to address this problem; methods include

- Artificial neural networks (Elsharkawy 1998; Abdel-Aal 2002; Osman and Al-Marhoun 2005; Alimadadi et al. 2011; Alarfaj et al. 2012; Alakbari 2016; Adeeyo 2016; Moussa et al. 2017; Ramirez et al. 2017; Arief et al. 2017)
- Support vector machines (El-Sebakhy et al. 2007; Anifowose et al. 2011)
- Nonparametric regression (McCain et al. 1998; Valkó and McCain 2003)
- Kriging and radial basis functions (Møller et al. 2018)

Nonparametric regression differs from parametric regression in that the shape of the functional relationships between the response (dependent) and the

	Predicted bubblepoint pressure	
	ARE, %	AARE, %
McCain et al. (Eqs. 7–12) (1998)	3.5	12.4
Velarde et al. (1999)	1.2	12.5
Labedi (1990)	0.0	12.6
Standing* (1947)	–2.1	12.7
Lasater** (1958)	–1.3	13.3
Levitan and Murtha (1999)	4.2	13.9
Al-Shammasi (1999)	–1.4	14.3
Vazquez and Beggs (1980)	7.1	14.6
Omar and Todd* (1993)	5.4	15.5
De Ghetto et al. (1994)	8.6	15.6
Kartoatmodjo and Schmidt (1994)	4.4	15.7
Dindoruk and Christman* (2001)	0.9	16.1
Glaso (1980)	4.8	16.8
Fashad et al.* (1996)	–5.6	17.8
Al-Marhoun* (1988)	8.8	17.8
Dokla and Osman* (1992)	0.3	21.8
Almehaideb* (1997)	–0.6	22.3
Khairy et al.* (1998)	4.9	23.1
Macary and El Batanoney* (1992)	12.6	23.1
Hanafy et al.* (1997)	10.6	28.8
Petrosky and Farshad* (1998)	17.7	37.7
Yi (2000)	42.4	45.2

*Author restricted the correlation to a specific geographical area.

**Not valid for °API<18.

Fig. 4.2—Comparison of published bubblepoint-pressure correlations (Valkó and McCain 2003).

explanatory (independent) variables is not predetermined but can be adjusted to capture unusual or unexpected features of the data. Fig. 4.3 shows the nonparametric regression results for bubblepoint pressures (AARE ≈10.9%), which compare favorably with measured quantities (Valkó and McCain 2003). This method circumvents the lack of measurement of stock-tank gas rate and specific gravity and estimates useful PVT properties from routinely available field operations data. Similar improvements were reported for solution GOR (AARE ≈ 5.2%) and gas gravity (AARE ≈ 2.2%).

Other recent authors have extended the effort to propose a stochastic description of the fluid (Alboudwarej and Sheffield 2016) for PVT uncertainty assessment.

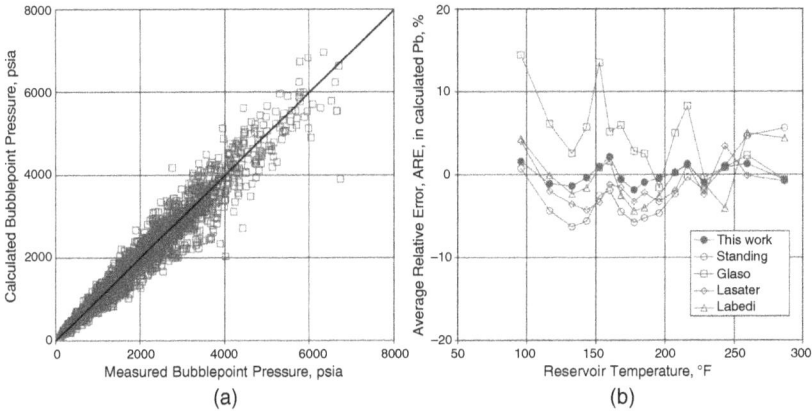

Fig. 4.3—(a) Calculated bubblepoint pressures compared with measured bubblepoint pressure for 1,745 data records, (b) AARE for calculated bubblepoint pressures compared with other correlations (Valkó and McCain 2003).

In unconventional assets, several operators use data-driven models sometimes combined with equation-of-state (EOS) modeling to establish a correlation between readily available field data, such as oil gravity or gas chromatography, and other derived fluid properties such as formation volume factors, viscosity, or saturation pressure (Yang et al. 2014; Yang et al. 2019a, 2019b).

Several authors present comparable or better AARE estimates of PVT properties using data-driven approaches for the fluid samples considered. In general, they consider routinely measured fluid properties from the field (e.g., GOR, pressure, temperature, API gravity, gas gravity) and train a supervised model (neural network or support vector machine) during the learning phase. The forward model is then validated and used to predict against the test data set. In contrast, the nonparametric regression method (based on alternating conditional expectation) establishes a nonlinear transformation on the input and output variables to discover the functional forms determining their relationship.

It must be noted that most of the published work listed above does not publish the forward model that can be independently verified by a practitioner on new data sets. In these cases, the results should only be considered as an investigation of the data-driven methods as applied to the specific fluid data set used in the study. In cases where these models are available, they must be tested on a blind data set with careful consideration of the assumptions and validity range before being used.

Supervised learning techniques for PVT estimation, though promising, are dependent on the availability of adequate and representative data sets. Therefore, when limited fluid samples are available within a field, analog field data could be carefully selected to augment the data set for better results.

In cases where large sample size of PVT data sets is available (such as unconventional or mature fields), better results might be obtained by training the regression or machine learning models on field-specific data. Wherever possible, one must examine the possibility of using a common EOS model across these data sets.

Summary

- *Data-driven modeling techniques extend the practice of estimating fluid properties through empirical correlations through a systematic way of deriving relationships from data.*
- *Supervised learning techniques can be used when sufficient field-specific PVT data are available.*
- *Published data-driven models for estimating PVT properties should not be used without checking on the assumptions and population statistics of underlying data used for training (e.g., ranges of GOR, API gravity).*

4.2. Core Analysis. Core data represent an important input in any reservoir model. Facies, porosity, permeability, relative permeability, and capillary pressure are among the most common parameters that are extracted from routine or standard core analysis. Due to the relatively high cost of core collection and analysis, data sets are usually relatively sparse. A common practice is to establish a correlation between core and log data that can then be applied more generally across the field. Neural networks have been used with great success to that end for more than 20 years (Mohaghegh et al. 1995) but the industry has recently seen a renewed interest in applying newer machine-learning algorithms to improve the correlations and extend the approach to more-challenging geological environments.

Mohaghegh et al. (1995) published the first application of artificial neural networks (ANNs) to correlate petrophysical logs and core properties. **Fig. 4.4** presents

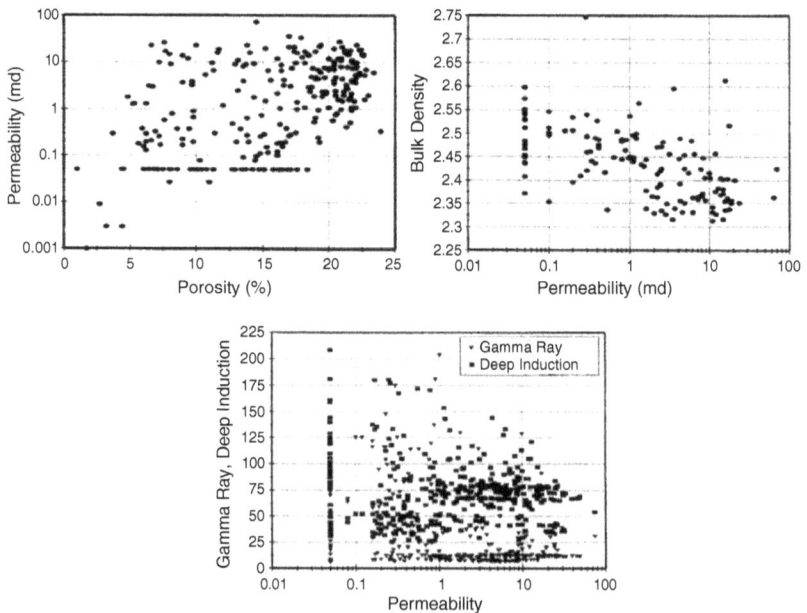

Fig. 4.4—Core permeability vs. petrophysical log properties.

the crossplots between permeability values determined from core analysis and the corresponding bulk density, gamma ray, and deep induction properties determined from wireline logs. Although one would be hard pressed to visually establish a correlation, an ANN applied to the problem was able to identify relevant patterns in the data set and provided a robust correlation that could be used as an artificial petrophysical log of permeability. The machine-learning-based permeability was compared to the core permeability on blind test samples and showed an excellent correlation ($R^2 = 0.963$). **Fig. 4.5** shows the comparison between the core permeability and the model based on the ANN.

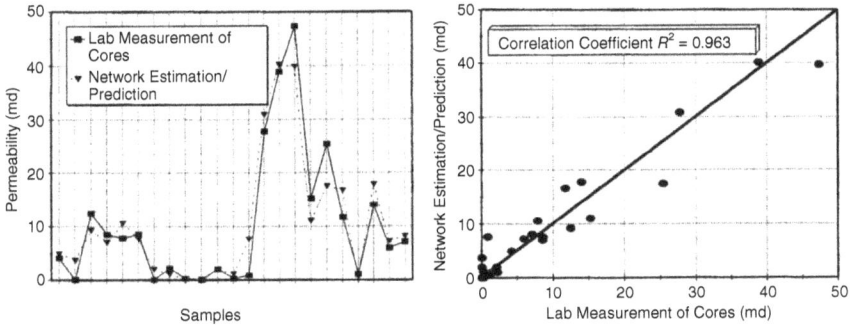

Fig. 4.5—Match between predicted and measured permeabilities.

Several authors have investigated the use of different machine learning methods for this problem and have documented their performance. Al-Anazi and Gates (2010) and Shahab et al. (2016) have, for example, reported the strong performance of algorithms based on support vector machine over different types of neural networks for this problem.

The approach of predicting core data from log curves has also been extended to other data sets. For example, Negara et al. (2016) published a workflow using support vector regression to correlate total organic carbon (TOC) obtained by core measurements to a suite of well log data (e.g., gamma ray, acoustic, resistivity, bulk density, and elemental spectroscopy).

Capillary pressures represent another type of data typically derived from core measurements that has been modeled using machine learning models. Mohammadmoradi and Kantzas (2018), for example, presented a study where they used an ANN to establish a correlation able to predict contact angle from the concentration of calcite, clay, quartz, and total organic carbon. Their work is focused on understanding the wettability of unconventional reservoirs where imbibition might play a significant role in fracturing-fluid uptake in the reservoir rock.

In a variety of petrophysical analysis efforts, machine learning algorithms are used to accelerate the manual interpretive work performed by experts. Many authors have documented workflows where large amounts of data were collected and a small fraction was analyzed manually by experts to establish a training data set. A machine-learning algorithm is then calibrated on the training set and used on the rest of the data to complete the analysis. Sidahmed et al. (2017) used a deep learning algorithm to perform automated rock facies classification from a standard petrophysical log suite. Hoeink and Zambrano (2017) used logistic regression to determine shale lithology from log data. Tang and Spikes (2017) used a neural network to perform image interpretation of scanning-electron-microscopy data, segment the images, and determine the mineralogy of different features. Bestagini et al. (2017) used a gradient boosting classifier in automated facies identification from well log data. Budennyy et al. (2017) used decisions trees with image preprocessing, k-means clustering, and principal-component analysis to perform rock type and mineral composition classification on a data set of petrographic thin-sections.

However, laboratory estimation of core properties for unconventionals with very low permeability is still a challenge and remains both a problem and an opportunity for data analytics applications.

Summary

- *Data-driven modeling techniques can be useful in estimating rock properties and identifying facies, when relevant training data are available.*
- *Upon proper calibration, a common application is to predict (infrequently available) core properties from (more commonly available) log data.*

4.3. Reserves and Production Forecasting. The economic success of reservoir exploitation relies heavily on resource estimations and production forecasts. The quality of these forecasts often defines the success or failure of many projects (Gupta et al. 2016). To provide reliable estimates of future production rates, all available geological and engineering data should be integrated. To account for the uncertainties surrounding these estimations, the evaluation of oil and gas resources, expected production, and reserves has been transitioning from deterministic to probabilistic. Probabilistic methods offer the advantage of capturing the variability of geological or engineering factors and help quantify the uncertainty ranges associated with the estimates.

4.3.1. Resource and Reserves Calculations. Empirical oil-recovery correlations were initially developed by the American Petroleum Institute (API) on sandstone reservoirs in the United States. Arps (1945) later adapted these methods by developing decline curve analysis (DCA). Since then, simple oil-production forecast equations have continuously been developed and the development of data analytics methods has recently revived these efforts. Correlations have for example been proposed to estimate the well's peak rate and expected ultimate recovery (EUR) using information such as hydrocarbon content, depth, well length, and thickness. These methods offer efficient well-by-well analysis for scenarios where it is not practical to perform DCA on each individual well (Sharma et al. 2013). Other authors have used multiple linear regression models to estimate the recovery of unconventional wells using well-design parameters (Cunningham et al. 2012). Advanced analytical models have also been used when richer data sets were available. For instance, ANNs have been used to estimate EUR from rock properties, well characteristics, and completion parameters (Javadi and Mohaghegh 2015). They have also been used to target production zones using lithological attributes (Da Silva and Marfurt 2012).

Data analytics applications are not limited to production forecasting of existing wells or assets. Some models have been used to estimate the recovery of a field prior to its development, using exploration and appraisal data sets.

Sometimes analytics models are used in conjunction with physics-based models, such as reservoir simulation, and are used to create surrogate models or proxies.

4.3.1.1. Data-Driven Expected Ultimate Recovery Prediction Methods. The EUR is a key driver for the economic viability of an exploration and production (E&P) asset, but EUR estimates are often associated with significant uncertainties (Lin et al. 2012) especially during the exploration and appraisal phases of a project.

To improve the early quantification of EUR, a data-driven methodology was proposed to estimate the variance in EUR estimated during the appraisal phase. A multivariate statistical method (partial least-squares regression) and optimization procedure was used to correlate known EUR values calculated during appraisal to data describing the characteristics, complexity, and definition level of 50 deepwater Gulf of Mexico oil fields (Gupta et al. 2016). The approach delivers a robust EUR estimation (R^2 = 73% on blind data sets) and identifies the key drivers impacting the variance. With improved feature selection through dimensionless numbers, the accuracy can be further improved (Srivastava et al. 2016).

4.3.1.2. Reservoir Analog Selection. Reservoir analogs provide valuable information to support development decisions. They are often used to evaluate development scenarios, estimate recovery factors, or evaluate the potential economic viability of development projects (Dursun and Temizel 2013). Data analytics applications have recently been proposed to improve the identification of analogs.

Case-based reasoning (CBR) has recently been used to identify reservoir analogs using both continuous and categorical attributes integrated using distance metrics and similarity measures (Dursun and Temizel 2013).

A common challenge in selecting analogs is the treatment of missing information, which introduces uncertainties. It is therefore important for analog selection methodologies to characterize these uncertainties and account for them when estimating the field production potential by estimating an associated risk. Various data-driven approaches have been proposed to that end, and their performance was tested against analogs selected by experts (Perez-Valiente et al. 2014). Decision trees were found to be the most effective algorithm for categorical properties (**Fig. 4.6**), while regressive support vector machines produced the best results for numerical properties (Perez-Valiente et al. 2014).

The proposed methodology shows how data analytics can be used to identify analogs for newly discovered reservoirs with limited information. An important additional outcome is the estimation of unknown properties such as recovery factor and production mechanism, and their associated uncertainties.

4.3.1.3. Recovery Factor Benchmarking. A key objective of a reservoir recovery factor benchmarking study is to support decision making on an ongoing development, well before the entire program has been executed. Possible applications include early confirmation of successful well placement, early indication of the impact on well performance because of changes to drilling and stimulation procedures, and a

CATEGORICAL PARAMETER	KNN	Neural Networks	Decision Trees	Support Vector Machines
BASIN CODE (KLEMME)	0.64	0.61	0.73	**0.75**
BASIN CODE (BALLY)	0.73	0.63	**0.80**	0.78
PRESENT TECTONICS CODE	0.73	0.66	0.73	**0.74**
FLUID TYPE CODE	**0.89**	0.77	**0.89**	0.86
PRINCIPAL STRUCTURAL CODE	0.61	0.54	0.57	**0.70**
PRIMARY TRAP TYPE CODE	0.69	0.60	**0.71**	0.70
SECONDARY TRAP TYPE CODE	0.35	0.37	0.44	**0.45**
NUMBER OF STRUCTURAL COMPARTMENTS	0.69	0.64	**0.72**	0.71
NUMBER OF STRATIGRAPHIC COMPARTMENTS	0.42	0.42	**0.52**	**0.52**
PRIMARY LITHOLOGY CODE	0.72	0.64	**0.78**	0.68
PRIMARY SEDIMENTARY SYSTEM CODE	**1.00**	0.96	**1.00**	0.95
PRIMARY SEDIMENTARY ENVIRONMENT CODE	**1.00**	0.96	**1.00**	0.93
DIAGENETIC PROCESS CODE	**0.89**	0.77	0.88	0.84
FRACTURED RESERVOIR CLASSIFICATION CODE	**0.80**	0.77	**0.80**	0.79
PRIMARY POROSITY TYPE CODE	**0.64**	0.48	0.57	0.53
SECONDARY POROSITY TYPE CODE	0.42	0.30	**0.45**	0.43
PRIMARY DRIVE MECHANISM CODE	**0.62**	0.52	0.58	0.59

NUMERICAL PARAMETERS	KNN	Multiple Linear Regression	Neural Networks	Decision Trees	Support Vector Machines
DIP ANGLE (°)	75.83%	83.63%	115.01%	81.38%	**74.62%**
AREA (Km²)	292.84%	228.88%	358.25%	300.81%	**186.38%**
ORIGINAL TOTAL HC COLUMN HEIGHT (m)	64.27%	54.91%	93.18%	74.5%	**52.65%**
TOP RESERVOIR DEPTH (m)	22.01%	12.98%	18.96%	21.59%	**10.61%**
ORIGINAL TEMPERATURE GRADIENT (°C/m)	26.11%	12.98%	10.33%	25.31%	**8.42%**
TEMPERATURE AT TOP RESERVOIR DEPTH (°C)	17.71%	16.31%	13.57%	20.98%	**8.08%**
ORIGINAL PRESSURE GRADIENT (KPa/m)	10.94%	7.81%	8.83%	12.21%	**5.79%**
PRESSURE AT TOP RESERVOIR DEPTH (KPa)	14.39%	9.01%	11.19%	15.45%	**5.64%**
AVERAGE GROSS THICKNESS (m)	69.11%	64.57%	128.79%	76.62%	**63.59%**
AVERAGE NET PAY (m)	73.21%	68.63%	101.73%	69.14%	**60.56%**
AVERAGE MATRIX POROSITY (%)	61.49%	63.35%	65.77%	**45.5%**	53.3%
AVERAGE AIR PERMEABILITY (mD)	783.06%	881.48%	1062.66%	671.25%	**429.14%**
AVERAGE WATER SATURATION (%)	77.15%	66.51%	68.43%	65.34%	**58.33%**
AVERAGE API GRAVITY	21.09%	23.6%	24.49%	19.4%	**18.97%**

Fig. 4.6—Performance machine learning algorithms for categorical and numerical parameters (Perez-Valiente et al. 2014).

conditional probabilistic outlook of long-term well behavior to better define well/field economic scenarios and to guide reserves bookings.

Various authors have proposed benchmarking methods to enhance the forecasting of long-term reservoir and well recovery performance, based on metrics of short-term well behavior (Van Den Bosch and Paiva 2012).

The typical metrics include measured peak production rates and cumulative volumes at various times. The process is refined continuously as new information becomes available, and the added value of the information is quantified.

A performance indicator summarizes the range of predicted outcomes into a single number quantifying the degree of misclassification, from zero, describing no predictive capability, to 100, describing a perfect prediction. This approach offers a transparent and unbiased benchmarking of the system's prediction capability for each metric (Van Den Bosch and Paiva 2012).

The lessons learned from a global analog reservoir knowledge base can be used to identify opportunities for reserves growth of mature fields through benchmarking (Lu et al. 2016).

A probabilistic method using a response-surface model was implemented to assess hydrocarbons in place, EUR, and recovery factor for shale gas reservoirs such as the Marcellus Shale (Richardson et al. 2016). **Fig. 4.7** shows a 25-year EUR response-surface model created using a semianalytical model. The presented analytical expression of EUR is a linear combination of key reservoir parameters. Both in-place volumes and ultimate-recovery models were coupled during Monte Carlo simulation to produce a probability distribution for the recovery factor.

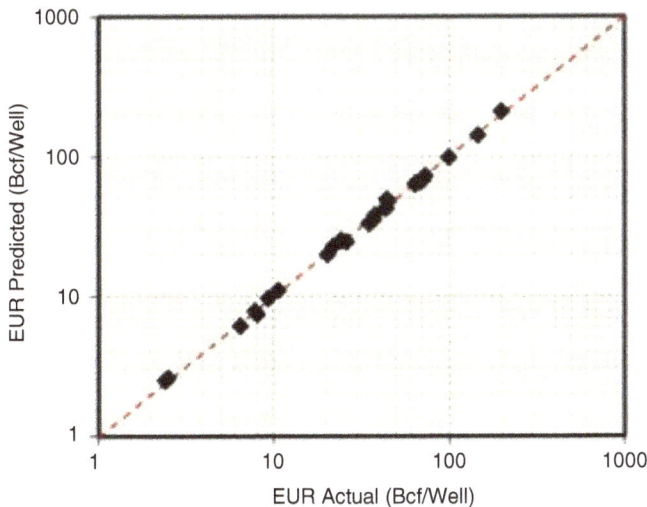

Fig. 4.7—Comparison of EUR predicted from a response-surface model and that from a semi-analytical model.

4.3.1.4. EUR Assessment Using Machine Learning Techniques. Numerical simulation offers nonunique solutions on history-matched calibrated models. Such a model-solutions ensemble has a direct impact on business decision making such as EUR assessment and reservoir drainage optimization. Multirealization history-matching techniques have been applied to quantify the EUR uncertainty. Because of the uncertainty associated with calibrated models and computational intensity requirements, machine learning techniques have been implemented to facilitate the investigation and further reduce reservoir performance uncertainty. The uncertain parameters typically include reservoir static properties, hydraulic-fracture properties, and parameters defining dynamic properties such as relative permeabilities.

As an example, a method to assess EUR for several wells in the Permian liquid-rich shale reservoir field (Guo et al. 2017) has been proposed. Each realization of the initial ensemble was calibrated iteratively using a distributed Gauss-Newton (DGN) method. The responses generated during iterations are added to a training data set, which was then used to train an ensemble of support vector regression (SVR) models. The sensitivity matrix for each realization is estimated analytically from the SVR models and used by the DGN method to generate improved search points and accelerate convergence. The integration of SVR into the DGN method allowed 65% of the simulation runs to be saved compared to the traditional DGN without SVR. This increased efficiency comes from the use of machine learning methods that continuously integrate the simulated results from the previous iterations. This is an example of how data analytics methods can be used in support of numerical simulations to provide faster EUR forecast and uncertainty ranges.

4.3.2. Production Forecasting. Today, engineers can choose from a range of production forecasting techniques. Although numerical simulation still plays a critical role in providing forecasts of hydrocarbon production, some data analytics techniques have exposed a new world of possibilities and insights not previously revealed by physics-based methods. Commonly, reservoir engineers use field measurements for reservoir model calibration, so their predictions agree with historical production. Calibrated models are then used to forecast future production to identify optimal production strategies that are resilient to reservoir uncertainty. The objective of including data-driven analytics in the forecasting process is to guide the investigation of reservoir scenarios and production strategies. It is important to rapidly understand how simulation results change with model parameters controlling the reservoir description and operating conditions.

"All forecasts are predictions, as we can use a model to simulate the future; but not all predictions are forecasts, as when we would use regression to explain the relationship between two variables or using a model to simulate the past." Forecasting requires logic, an ability for quality assessment of forecasting approaches, and few rules for effective forecasting.

We should consider forecasts to allow the decision maker to do, for example, the following:

- Exercise strategic judgment
- Identify key patterns and seasonality
- Embrace those items which cannot be classified
- Look at more past or historical data to make sense when fewer data elements would not make any meaningful forecasting

We also need to understand when to make a combination of forecasts or forecasting methods by using ensembles and when not to forecast at all. Forecasting and selecting an appropriate method for performing forecasting will always be an interesting blend of art and science in addition to our judgment and practicality (Mishra 2017).

4.3.2.1. Analytical Models. Using the traditional reservoir body of knowledge, engineers estimate theoretically expected well productivity—and, hence, predict reservoir flow rates—using approximations for reservoir dimensions, energy, rock and fluid properties, and the impact of reservoir exploitation decisions including drainage architecture and recovery methods. In addition, production measurements and analog reservoir response have the power to enhance the confidence of these estimations.

Traditional analytic reservoir engineering methods include material balance and production rate analysis. Most of these methods are derived from first-principles models, simplified using empirical observations and calibrated to direct measurements including direct measurements of production rate and pressure data. The two most common analytical models used for production forecasting are decline curves and type curves.

Some of the oldest and most frequently used data-driven methods for production forecasting are those related to decline curves, including harmonic, hyperbolic, and (the most common) exponential (Arps 1945). The popularity of these methods derives from their simplicity: A few parameters are sufficient to calibrate the model and provide a forecast of the future well production behavior.

Decline curves have been used in many instances for production forecasting at the well and the reservoir level. They can be used to determine expected remaining recovery and are quite useful to describe the typical behavior of wells by determining a representative decline curve. DCA is the most common method used to estimate reserves and resources.

However, the Arps method is not directly applicable to unconventional reservoirs and would lead to significant overestimation of reserves. Newer methods such as power-law exponential decline (Ilk et al. 2008), stretched-exponential decline (Valkó 2009), Duong (2010), logistic growth model (Clark et al. 2011), and others (Artus et al. 2019) have been proposed in the form of empirical equations with a few fit parameters to describe observed decline behavior (Ali and Sheng 2015).

The use of decline models, however, is limited to estimate production behavior for known operating conditions and is inappropriate for optimizing reservoir management strategy in terms of well location, wellhead pressure control, or number of wells.

Type curves are powerful graphical representations of the theoretical solutions to transient and pseudosteady-state flow equations. Reservoir engineers are usually challenged to find a match between historical reservoir performance (e.g., rates and pressure) and a theoretical type curve (Agarwal et al. 1970; Fetkovich 1980; Carter 1985; Palacio and Blasingame 1993). Type curves are usually represented in terms of dimensionless variables, including dimensionless pressure, rate, cumulative production, time, radius, and wellbore storage.

Matching becomes a bit like an art instead of science because real data can be noisy and might contain outliers. In addition, actual reservoir architecture might not exactly fit the assumption made for the available models.

In this sense, type-curve analysis becomes a key area where machine learning and AI can be used to augment engineers' knowledge. Pattern recognition and CBR have been used to derive the parameters of type curves (Saputelli et al. 2015).

4.3.2.2. Reservoir Simulation Proxies. Engineers routinely use reservoir simulation models to forecast reservoir and well production. The physics-based models are calibrated against field data and are frequently used to support field development and other reservoir management decisions.

Analyzing complex reservoir simulation models is challenging. The standard approach that consists of visualizing saturation and pressure fields or production profiles is time-consuming and often fails to clearly establish the relationship between the input and response parameters.

Proxy models approximate the simulator response parameters from the inputs. A proxy is a simplified analytical model or machine learning model used to describe a physical phenomenon within a parameter space. These models can be as simple as polynomial regressions of field production rates subject to changes in operational conditions or reservoir parameter uncertainties. They can also be as complex as emulating the entire output from a reservoir simulator, including the pressure and saturation results in each grid cell. These simpler models are often referred to as response-surface models, and the more-advanced models are sometimes called surrogate models.

These models have been applied to a range of problems, from steam-assisted-gravity-drainage (SAGD) processes (Queipo et al. 2001) to well target optimization (Yeten et al. 2005). The choice of experimental design and response-surface methodologies has an impact on the outcome. **Fig. 4.8** shows the comparison of estimation errors among multiple response-surface methods for various experimental design choices for a reservoir case to predict oil production curves from simulation experiments. The authors conclude that the central composite design performs better than similar density D-optimal design, while neural network performed poorly in most cases.

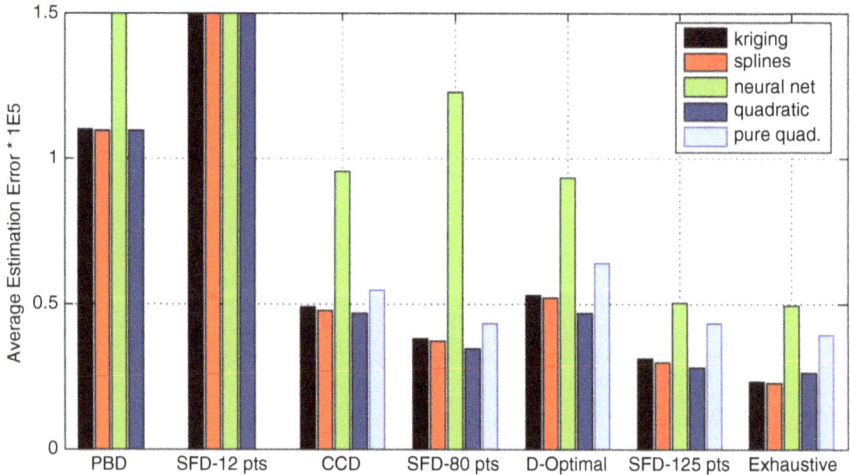

Fig. 4.8—Comparison of estimation errors among various experimental design and response-surface methods (Yeten et al. 2005).

Learning from physics-based models offers two key advantages. First, a broad and high-population data set can be generated to help train the surrogate models. Second, they can be made to properly learn the required physical relationship at play. In addition, when trained these models can be interrogated in a matter of seconds as opposed to hours for reservoir simulation models (Mohaghegh et al. 2006).

Surrogate models can be applied to coupling with the surface model in integrated asset models, to real-time optimization, to real-time decision making, and to analysis under uncertain conditions.

4.3.2.3. Reduced-Physics Models. Fluid flow models can be based on first principles (e.g., conservation of momentum, mass, and energy), empiricism, or a combination of both. First principles can be combined with constitutive equations to generate models that are valid over a wide range of operating conditions. However, they might be cumbersome to develop and manipulate. Empirical models, on the other hand, can be easy to develop, but might not be accurate outside the range of data used for their calibration. Reduced-physics models (described in Section 3.1) combine first principles with empirical constitutive equations (e.g. Darcy's law, ideal-gas law). These models are often easier to develop and manipulate than raw first-principles models and maintain their applicability outside of the range of data used for their calibration.

Engineers for example use reservoir simulation to identify the location and size of unswept regions, to quantify the degree of communication between injectors and producers, and to estimate the recovery efficiency in a region of the reservoir. These insights allow engineers to propose changes to reservoir management strategies designed to optimize the reservoir performance.

These models rely on a multivariate reservoir model to represent the variation in well and reservoir behavior in time and space, and they often combine data-driven and physics-based elements. Several multivariate reservoir modeling techniques have been published. Here we discuss a recent model that has gained significant traction over the past decade or so: namely, CRMs. CRMs get their name from an analogy between fluid flow in porous media and current flow in an electrical system (Bruce 1943). The derivation of CRM is based on enforcing mass balance on the drainage volumes of producing wells and can account for the influence of nearby injectors and the changes in the well operating conditions. These models solve equations similar to those that are used in reservoir simulation, but instead of using the reservoir pressure, the models directly estimate the well rates. This approach transforms the set of second-order partial-differential equations usually solved in reservoir simulators into a first-order ordinary-differential equation with an analytical solution. This simple method is fast enough to be used in reservoirs with high well counts and has been applied to water- and CO_2-injection problems (Albertoni and Lake 2003; Yousef 2006; Sayarpour 2009a, 2009b; Weber 2009; Salazar-Bustamante et al. 2012; Holanda et al. 2015; Gladkov et al. 2017).

In effect, the CRM is an extension of the exponential decline curve model that accounts for changes in operational conditions and for the influence of injectors. The pressure effects are estimated using the CRM equation. The saturation effects are usually modeled using empirical fractional flow models. The model offers a prediction of the well performance that is based on its historical decline and its production response from nearby injection that is quantified using connectivity factors with nearby injectors and parameters for an empirical fractional flow model. **Fig. 4.9** compares the performance of CRMs against other traditional reservoir engineering methods such as the empirical power-law fractional flow model (EPLFFM) and a Buckley-Leverett-based fractional flow model (BLBFFM).

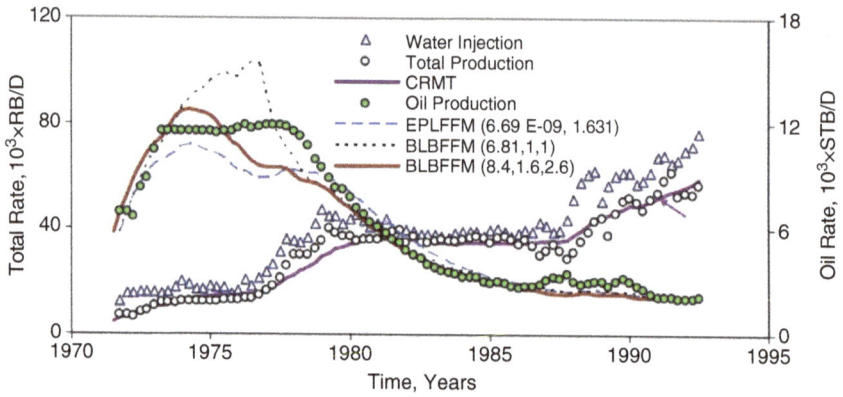

Fig. 4.9—Overall performance match with different fractional flow models, Reincecke Reservoir (Sayarpour et al. 2009a).

Compared to reservoir simulation, these models offer the advantage of being extremely fast to run and calibrate. Compared to simpler analytical models, they account for critical aspects of the reservoir management strategy and can be used effectively to support field optimization decisions. Compared with purely data-driven models, they offer the advantage of being based on physical relationships, which allows them to be used with confidence for moderate extrapolation.

More recently, Molinari et al. (2019a, 2019b) illustrated a hybrid data- and physics-based modeling approach that uses the diffusive-time-of-flight (DTOF) concept to estimate productivity-based production forecasts for unconventional or tight reservoirs. The method uses the DTOF and depth of investigation concepts to calculate drainage volume as a function of material balance time using only bottomhole pressure and rate data. By combining this with generalized material balance, the pressure depletion and productivity index evolution over time are computed, providing a productivity-based production forecast. This method is tested on several hundreds of wells and is shown to perform better than a standard rate-decline-based forecast (**Fig. 4.10**), especially under

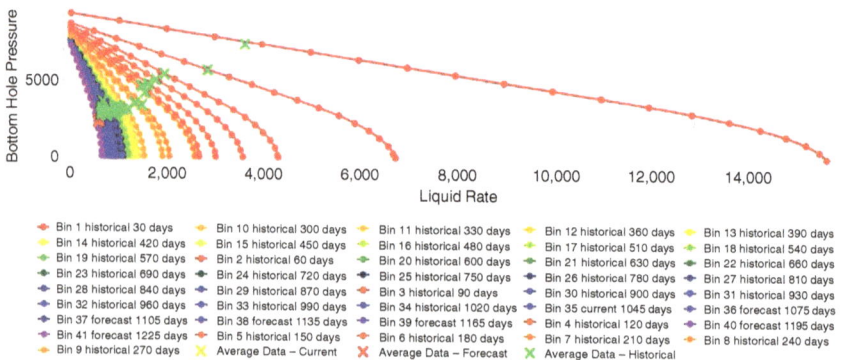

Fig. 4.10—Dynamic inflow-performance-relationship curves for every monthly bin (historical, current, and forecast) (Molinari et al. 2019a).

varying surface conditions (i.e., changes in tubing pressure control). The method also serves to be more practical than analytical or semi-numerical model-based approaches that cannot be scaled to every well in the field.

4.3.2.4. Data-Driven Models. Top-down data-driven reservoir models are data-driven models directly calibrated against field data (Gomez et al. 2009). Their key advantage is their complete flexibility in learning patterns that are challenging or time-consuming to model. Once trained, they can be interrogated almost instantly, which is ideally suited for optimization or sensitivity studies. The downside of top-down data-driven reservoir models is that they require an extensive data set to be trained and because they do not have embedded physical relations, extrapolation should be performed with caution.

These models can forecast reservoir and well performance in time and space with a certain degree of confidence and granularity. They can incorporate a broad range of data (pressure, temperature, well logs, core data, well flow tests, raw and interpreted seismic, and production/injection history), and account for both development activities (well count, location and trajectories, stimulation jobs) and the operational decisions (e.g., fluids injected, well targets, choke settings).

4.3.2.5. Statistical Time-Series Forecasting. Purely statistical methods are sometimes used to forecast time-series data. This approach allows time-dependent parameters to be predicted without the need to assume an underlying physical model. Common applications of this approach include the forecasting of rates and pressures at the reservoir or well level; production ratios, such as GOR or water cut; well productivity index; or flowing wellhead or downhole pressure or temperature.

Most time series are described using two basic components: trend and seasonality. The trend represents the general evolution over time. The seasonality is a recurrent pattern that occurs at various frequencies—for example, daily or seasonal temperature variations. Models often include an additional noise component, usually centered on the trailing average. Several approaches exist for such problems that are well-documented in the statistics and financial modeling literature (Box and Jenkins 2008).

As with all statistical models, a good fit to the data does not necessarily indicate a robust forecast. With enough parameters, a very close fit can usually be obtained but such a model would often fail to identify the key pattern in a time series and would offer a poor forecast. To alleviate this issue, a validation strategy should be used through data partitioning. A blind test can thus be developed to quantify the predictive power of a model.

These models should be used only to forecast data within a stationary range (i.e., within a period where the data are statistically constant). This situation is met in practice for pseudosteady-state conditions or relatively short production periods, past which the forecast should lose reliability.

Statistical forecasts offer the advantage to produce a confidence interval around the base prediction. Understanding the upper and lower expectations of a forecast might be of high interest in practice for some applications (to estimate the required name-plate capacity for facilities, for example).

Summary

- *Data analytics has revived and overcome the limitations of traditional fore-casting methods (e.g., DCA, analogs, type curves) that might not be readily applied outside of their strong assumptions.*
- *Reduced-physics or hybrid models offer the advantage of being extremely fast to run and calibrate compared to numerical methods, while still retaining the accuracy and ability for moderate extrapolation.*
- *Pure data-driven models need to be used with care for forecasting within stationary ranges based on the training population statistics.*

4.4. Reservoir Surveillance and Management. Reservoir management is concerned with maximizing the productivity and profitability of oil and gas fields throughout the full life of the asset, from selecting the right development strategy to optimizing a maturing field all the way to identifying the right divestiture or abandonment option.

Reservoir management is therefore a decision-making discipline. These decisions are often supported by a variety of models. Depending on the application, the models can be simple such as heuristics-based correlations, analytical solutions, or simplified physical representations, or can be as complex as advanced machine-learning algorithms or full-physics numerical models.

Every model used for decision making has a specific data requirement for its calibration. Reservoir surveillance is the tactical effort designed to collect the right data to support an effective reservoir management strategy. An optimal surveillance strategy maximizes the amount of information collected while minimizing the associated cost. Surveillance plans are usually concerned with measuring the nature or current state of the field.

Over time, the continuous reduction of sensor costs and the ever-expanding capabilities of data analytics methods have enabled significant improvements in reservoir surveillance and reservoir management efforts. This section presents a few published examples showing how data analytics has been used to improve reservoir surveillance and management.

4.4.1. Reservoir Surveillance. Capturing the right data is essential to enhance our understanding of the reservoir and associated wells and facilities and to support the critical reservoir management decisions to be made. As the reservoir matures, the type and frequency of data that needs to be collected evolve. Early in the field life, surveillance is usually focused on reducing subsurface uncertainties. Later on, data acquisition tends to focus on monitoring wells and surface facilities for optimization purposes.

The data acquisition frequency is dictated by the physical system being monitored. Static properties can be estimated using one-time measurements, but the dynamic system might require real-time monitoring at high frequencies (Alimonti and Falcone 2004), using measurement systems such as permanent downhole gauges for pressure, clamp-on acoustic sensing for sand, optical fiber for production allocation, or microseismic acquisition to monitor hydraulic fractures.

The reduction in sensor costs and the introduction of new measurement devices have increased the volume, frequency, and diversity of data that are being collected. Data analytics has helped reservoir surveillance in two general directions: the automation of routine analysis and the assimilation of complex data sets.

4.4.1.1. Automated Interpretation of Surveillance Data. Data analytics methods are sometimes used to accelerate or automate the analysis of surveillance data. Pressure- or rate-transient analysis (PTA/RTA) is a common approach used to analyze a variety of well and reservoir performance factors such as well productivity parameters, well interference, reservoir extension, and connectivity (Lee et al. 2003).

Permanent downhole gauges and multiphase flowmeters made high-frequency pressure and rate data available at all times and enabled the analysis of transient behavior for both planned and unplanned well shut-ins, which quickly translated to an increased workload for engineers.

New algorithms are actively being developed to automate such analysis. Sankaran et al. (2017) present a field application of a technology platform developed to analyze transient well behavior. They use a variety of algorithms to automatically analyze various well periods, such as shut-ins or ramp-ups, and apply a machine learning algorithm originally developed by Tian and Horne (2015) to perform the transient analysis. The kernel ridge regression method that was applied is a kernel method that uses a regularization term to avoid overfitting. The kernel method itself allows for a flexible mapping of the input parameters into nonlinear attributes and offers an efficient formulation in terms of computational efficiency. The method proposed proves to be much better than classical regression approaches in terms of both speed and robustness in the face of complex reservoir behavior. The methodology was used by Sankaran et al. (2017) to estimate productivity indices from permanent-downhole-gauge data, and **Fig. 4.11** shows the close match between measured and estimated productivity indices.

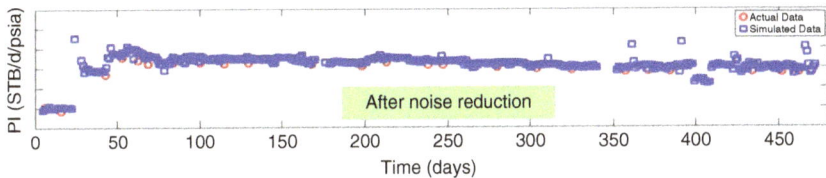

Fig. 4.11—Machine-learning-based well productivity estimation (Sankaran et al. 2017). PI = productivity index.

A significant amount of time has historically been spent by engineers locating the data required for analysis. Routine engineering analyses, such as PTA or decline curve matching are also time-consuming tasks that are subject to the bias of the individual performing the interpretation. Automating such work offers the dual advantage of ensuring consistency in the analysis and allowing engineers to spend more time on problems that require deeper analysis.

4.4.1.2. Surveillance Data Assimilation. Surveillance data should be assimilated as quickly as possible to maintain up-to-date reservoir models. The complexity of the reservoir model constrains how quickly new data can be integrated. Analytical models such as material balance or decline curves are fast enough to be updated almost instantaneously, but they rely on a simplified representation of the problem which often limits their applicability. More-general methods such as reservoir simulation can require months of work for an updated history match to be completed. For decades, this posed a challenge for large mature waterfloods. Today, several alternative data-driven models have been developed and used successfully, ranging from reduced-physics models such as streamline-based methods (Thiele and Batycky 2006), CRMs (Sayarpour 2008), or tracer-based approaches (Shahvali et al. 2012) all the way to fully data-driven models such as the neural networks used by Nikravesh et al. (1996). The next section will cover this topic in more detail.

In addition to being used to accelerate data assimilation efforts, data analytics methods have been leveraged to integrate complex data sets. Reservoir surveillance usually includes measurements as varied as seismic and microseismic surveys, well logs including image logs, cores, fiber-optic data, and pressure and production data (e.g., Raterman et al. 2017). Data-driven models offer a flexible way to account for eclectic sources of information that do not necessarily have a systematic framework for integration into standard physics-based models. Such an approach is used heavily in unconventional-reservoir modeling. Methods such as multivariate regression have been used to predict the performance of unconventional wells from varied data sets that could not be integrated into conventional physics-based models (Ciezobka et al. 2018; Burton et al. 2019). These models are used by unconventional-field operators for business planning and development decisions.

As new methods emerge for processing and analyzing real-time data, new data types (besides rate, pressure, and temperature) are also emerging that capture different physics (such as fiber-optic distributed temperature and acoustic sensing and tracer methods).

4.4.2. Reservoir Management. Common applications of data analytics to reservoir management have targeted two main objectives: the systematic identification of field development opportunities and the automation of reservoir optimization efforts.

4.4.2.1. Opportunity Identification and Ranking. Various actions can be taken on a given field to increase its recovery or financial value: for example, drilling new wells or sidetracks, optimizing or upgrading existing artificial lift systems, or starting or improving an injection strategy. Traditionally, it has been the role of the asset team to identify and prioritize these field improvement opportunities. The work is typically conducted by multidisciplinary teams using established workflows to analyze field data, apply heuristics, or create models to determine whether an action is feasible and financially accretive. Aging assets tend to grow in complexity, with an increasing well count, more- mature reservoir conditions, depleted pressures, increasing GORs and water cuts, and aging facilities and wellbores that can include various vintages of artificial lift systems or completion designs. Required maintenance operations are also multiplying, making it increasingly challenging to find the time to identify new opportunities.

Data-driven methods have been used successfully to automatically identify, categorize, quantify, and rank value-creation opportunities. The various heuristics and workflows traditionally used by asset teams can be programmatically implemented and applied systematically to all wells, thus methodically testing all known opportunity categories. Brown et al. (2017) present proven workflows previously developed through experience and applied manually that have been replaced by data-driven equivalents to identify field development opportunities. For example, a pay-behind-pipe identification algorithm was developed that automatically processes well logs to identify highly productive zones. Historical perforations are then analyzed with historical production data to automatically define the amount of oil that has been produced from each perforation over time. Remaining oil behind pipe is then estimated using a trained neural network, and potential reperforation opportunities are presented with a quantified recoverable oil volume and associated uncertainty. Each opportunity identification process that is being mapped allows for the work to be performed much faster and at a vastly different scale. Every process implemented saves time for engineers and geoscientists. Additionally, large work efforts that used to be split among different individuals in a team can now be processed through a single algorithm ensuring consistency and repeatability and reducing the risk of individual bias in the assessment.

Although each individual process is beneficial by itself, the transformative value gain comes from compounding a variety of processes together, so that a number of different opportunities can be compared and ranked against one another. The study presented by Brown et al. (2017) was performed on an 85-year-old field containing 800 wells and five stacked reservoirs. Out of that, 36 behind-pipe opportunities were identified and estimated to be above the economic hurdle of the operator. These were compared with other types of reservoir management opportunities such as new-drill locations, waterflood optimization and expansion, or artificial lift changes.

Data analytics offers numerous methods to tackle opportunity identification problems. Mohaghegh (2016) uses the top-down modeling terminology to describe a data-driven reservoir modeling approach that relies on machine learning and data mining techniques to identify new reservoir development opportunities. Kalantari Dahaghi and Mohaghegh (2009), for example, use fuzzy pattern recognition to identify optimal infill-well locations in the New Albany Shale. The methodology has been applied on many conventional and unconventional fields worldwide and has demonstrated that data analytics can be applied to quickly identify complex reservoir management opportunities in mature assets. A similar approach was also used by Sharma et al. (2017) to identify well intervention candidates in the Coalinga Field located in the San Joaquin Valley in California. The authors combined Voronoi grids and fuzzy logic to create an algorithm able to quantify the potential of well interventions in the mature heavy-oil reservoirs. The algorithm was applied to 1,700 wells, and the predictions were successfully blind tested for validation against recent well interventions (**Fig. 4.12**). The study was then used to identify and prioritize future opportunities that were successfully implemented in the field.

4.4.2.2. Real-Time Optimization. Data analytics has played an important role in a decades-long effort toward automating reservoir management and control to strive toward real-time optimization of producing oil and gas fields. The general framework used is presented next, followed by the concept of integrated asset modeling, which is one of the core components of digital-oilfield efforts.

**2013-2014 Workover Lookback:
CI Weighted Voronoi OIP vs. Actual Production**

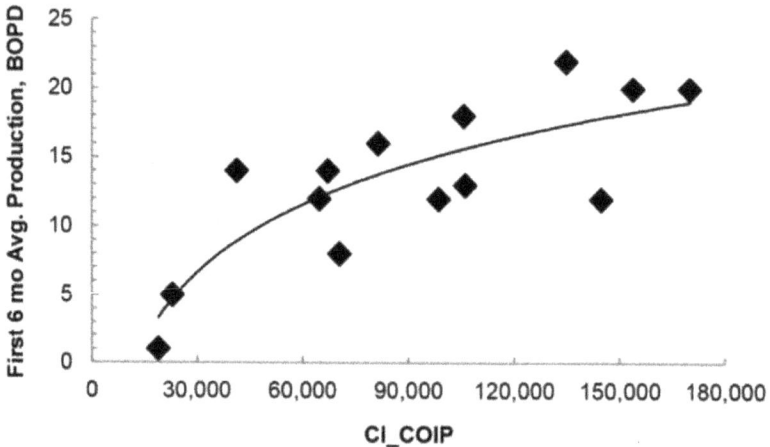

Fig. 4.12—Predicted (solid lines) vs. blind-tested (dots) well interventions (Sharma et al. 2017).

General Framework. The term real-time optimization is used to describe a process where data acquisition, decision making, and implementation actions are performed at the same frequency. In reservoir management, real-time optimization usually aims at finding the optimal operational conditions for the integrated system: reservoir, wells, and surface facilities. This is evidently an extremely challenging task because of the complexity of the infrastructure and the inherent uncertainty of the subsurface. This is further complicated by the wide range of time scales involved: blending decisions that are made for a multiyear field development with decisions made minute by minute by an automatic flow controller is a daunting task (Saputelli et al. 2000).

To tackle this challenge, the real-time optimization problem is usually decomposed into a hierarchy of levels (Sankaran et al. 2011), each level being associated with a different time scale. The optimization is then performed at each level with an appropriate frequency.

The feedback scheme encompasses a broad class of optimization and control paradigms, ranging from optimization for field development and production planning and scheduling, to second-by-second feedback control of flow rate through valve adjustment. For example, decision making at the upper levels might involve various optimization paradigms concentrating on explicitly stated economic objectives, whereas decision making at the lower levels might be automated and focus on engineering objectives, such as those by means of standard proportional-integral-derivative controllers.

Integrated Asset Modeling. Integrated asset modeling (IAM) is a mathematical representation of the production and injection system used to compute multiphase flow rates, pressure, and temperature throughout the field. IAM integrates different models representing the reservoirs, wells, pipelines, and gathering, processing, and export facilities. This contrasts with the traditionally isolated modeling which assumes that each individual model has fixed boundary conditions. For example, a reservoir simulation model might assume a constant bottomhole pressure constraint for its wells. IAM will instead pass information between a reservoir and a wellbore model at each timestep so that the output of one component becomes the input for the next one, forming a fully coupled system.

Surface network simulation technology was introduced by Startzman et al. (1977) and has gained traction in the last few decades to improve the accuracy and precision of production forecasts. In general, the success of IAM lies in integrating existing models, so that each component can be maintained as usual by the discipline specialists.

A key challenge for IAM has been to deliver acceptable run times and ensure stability. To tackle these issues, data-driven prediction and forecasting models (e.g., surrogates and/or proxies) have been used in the IAM context. These data analytics approaches are used to replicate the full physics at higher speeds.

Another practical challenge for IAM is to break the barriers and promote collaboration in several operating companies, as the teams responsible for various components of the IAM often have different cycle times for updating and managing the models. The ownership of the combined entities as part of the IAM is often not well-established.

Various data-driven simplifications of the subsurface response have been used in the context of IAM for driving complex asset management decision making. A DCA-based algorithm is used to rapidly determine the rig drilling schedule, more specifically to determine the investment timing for large offshore gas fields with the objective of sustaining nominated gas volumes (Aulia and Ibrahim 2018). Applications have been published in many types of fields. In coalbed methane for example, Shields et al. (2015) have successfully incorporated a predictive model in the form of pressure- and time-dependent type curves into a hydraulic model of the surface network to deliver an integrated production model.

Reservoir Management and the Digital Oil Field. A self-learning reservoir management strategy can be achieved by combining parametric fluid flow modeling, model predictive control, and economic (net present value) optimization for data-rich instrumented fields (Saputelli and Nikolaou 2003). Several industry case studies have demonstrated the value of digital-oilfield technologies for successful reservoir management (Adeyemi et al. 2008; Sankaran et al. 2009).

Sankaran et al. (2011) show how a digital-oilfield effort was used on the Agbami Field in Nigeria to deliver significant reservoir management benefits. Seven case studies are presented that highlight the impact of the digital-oilfield approach taken—e.g., improved zonal-crossflow management, workover risk mitigation, and tighter conformance control. The work effort created millions of US dollars of incremental value and allowed the operating company to get closer to management by exception by automating routine tasks.

Summary

- *Data analytics can improve reservoir surveillance in two general areas— automation of routine analysis and assimilation of complex data sets.*
- *Common applications of data analytics in reservoir management include identification of field development opportunities and automation of reservoir optimization.*

4.5. Enhanced Oil Recovery and Improved Oil Recovery. EOR is a method used to enhance hydrocarbon recovery through modification of physical and chemical properties of the rock and/or fluid. IOR covers a range of methods used to enhance hydrocarbon recovery from reservoirs that are not on primary production.

Data-driven analytics are a prime technology to be applied for EOR and IOR projects and operations because of the wealth of data offered by these operations (Kronkosky et al. 2017; Popa et al. 2011). In this section, we will cover a few applications of data analytics in that realm—namely, for IOR/EOR screening as well as for waterflooding and steamflooding operations.

4.5.1. Screening Tools for EOR/IOR. Selecting the best EOR/IOR method is a complex and time-consuming process involving significant amounts of data and a considerable number of simulation runs. Ultimately, an economic model should be coupled with the final production performance to assess the viability of the recovery technique.

When selecting the recovery method, the fundamental parameters included in any reservoir model are the rock, fluid, and formation properties because these are unique to each reservoir. The selected recovery process that would return the highest economic outcome determines the design parameters. Because the application of an EOR/IOR process usually follows the primary-recovery mechanism, a large amount of data and information has already been captured regarding the reservoir characterization and is available for use during screening.

Many approaches have been taken for IOR/EOR screening; however, this section will present only those that used intelligent data-driven analytics. The conventional methods are often driven by field analogies, pilot projects on a portion of the field, or prior operational experience on similar reservoirs. This approach could pose challenges such as lack of objective rules to define a reservoir type or the project completion time, or could have bias based on expert opinions.

Expert system-based approaches to EOR screening have been proposed (Guerillot 1988; Zerafat et al. 2011) that used an inference engine with an underlying knowledge base system. Such approaches can only consider technical criteria because the economic criteria can differ among geographical areas and companies. Other dimensionality-reduction techniques have also been proposed that are based on clustering and rule extraction algorithms (Alvarado et al. 2002) for screening EOR/IOR potential and have been applied to Venezuela mature fields.

More recently, a screening toolbox (Parada and Ertekin 2012) consisting of proxy models that implement a multilayer cascade feed-forward back-propagation ANN algorithm has been proposed for a diverse range of reservoir fluids and rock properties. The field development plan is featured in this tool by different well patterns, well spacing, and well operating conditions. The screening tool predicts oil production rate, cumulative oil production, and estimated production time for different sets of inputs, which facilitates comparison of various production strategies such as waterflooding, steam injection, and miscible injection of CO_2 and N_2. Drilling and completion techniques, well pattern, well spacing, and the recovery mechanism were used as design parameters. The ANN tool is able to recognize the strong correlation between the displacement mechanism and the reservoir characteristics, as they effectively forecast hydrocarbon production for different reservoirs. The blind tests performed show that the ANN-based screening tool is able to predict the expected reservoir performance within a few percent of error.

Other approaches have been reported with applications to field case studies such as Bayesian classification and feature selection (Afra and Tarrahi 2015), probabilistic principal-component analysis and Bayesian clustering (Siena et al. 2015; Tarrahi et al. 2015), neural networks (Surguchev and Li 2000; Okpere and Njoku 2014; Yalgin et al. 2018), and genetic algorithms (Armacanqui et al. 2017).

4.5.2. Waterflood Management. Waterflooding is the oldest and the most common IOR method used in the industry and is usually implemented following primary recovery. Waterflooding is designed to compensate for the pressure depletion in the reservoir and to displace incremental hydrocarbons. Waterfloods usually involve many producers and injectors organized in different patterns depending on the reservoir characteristics.

Waterflood optimization usually aims at maximizing the oil recovery per barrel of injected water under specified reservoir and surface constraints (Sudaryanto and Yortsos 2001). Given the amount of data usually available in waterflooding projects, an excellent opportunity for data-driven optimization techniques is often presented. Different methods are used for reservoir management, well placement, water shutoff, rate optimization, and performance forecasting, to name only a few objectives (Das et al. 2009; Lerlertpakdee et al. 2014).

Data analytics methods have been successfully applied to fieldwide waterflood management in numerous reservoirs around the world. Models were trained to forecast recovery and optimize the water injection and production targets. The application of data-driven techniques is of specific interest when the reservoir is too complex to be accurately modeled. This often occurs in complex geological settings but is also often a result of the number of wells involved or the extensive history of the field. Classical waterflooding optimization methods, such as basic pattern calculations or advanced reservoir simulation modeling, can be impractical for such projects.

A data-driven approach (Nikravesh et al. 1996) that takes advantage of the historical data from a large waterflood field is used to construct several neural networks, which correlate the individual-well performance behavior as a function of the well history itself and the injection/production conditions of the surrounding wells or pattern. The intelligent system consists of an ensemble of neural networks with different functions. Specialized neural networks accurately predict wellhead pressure as a function of injection rate, and vice versa, for all active injectors. However, the primary neural networks are trained to history match oil and water production on a well-by-well basis and predict future production on a quarterly or biannual basis. The global optimization allows for designing the water injection policies that lead to the minimum injected water and the highest oil recovery. The distinctive element of this data-driven approach involves the division of the waterflooding field into regions of similarly behaving wells, thus accounting for different reservoir properties and heterogeneities, and it captures the relationship between injection and production within each region. In addition to injection/production optimization, the system is also used for water-breakthrough time prediction, as well as infill drilling performance.

4.5.2.1. Well Placement Optimization. Another application of data analytics in waterfloods has been the optimization of well placement. Determining the optimal location of new wells is a complex problem that depends on rock and fluid properties, well and surface equipment specifications, field history, and economic criteria. Various approaches have been proposed for this problem. Among those, direct optimization using the simulator as the evaluation function, although accurate, is in most cases infeasible because of the number of simulation runs required.

Data-driven approaches offer new alternatives to address these types of optimization challenges by involving machine learning techniques such as hybridization of genetic algorithms and neural networks. An example of this approach is showcased in Güyagüler et al. (2000). In that article, a methodology is presented to optimize well placement in waterfloods and the approach is used to determine the optimal location of up to four water-injection wells in the Pompano Field in the Gulf of Mexico.

The hybrid algorithm involves a genetic algorithm and neural-network-based proxy function to mimic the role of the numerical simulator. A neural network was trained and validated with simulation results. An initial development plan (initial well population) is created and then run through the proxy (neural network) to provide the expected performance in the current well location. The net present value (NPV) of the waterflooding project was used as the objective function to calculate the optimized-injection-rate design. Ultimately, it was noted that this intelligent data analytics approach for optimization of well placement and pumping rate is very general and extremely fast. It also has the potential to identify solutions that might be unnoticed by more costly conventional techniques.

4.5.2.2. Water Shutoff. In mature waterfloods, specifically after water breakthrough, reducing the excess water production without hurting the oil production significantly improves operational costs and ultimately the final recovery. Water shutoff and conformance control represent not only a significant financial incentive, but also an environmental motivator for the oil industry.

Several techniques have been developed to address water shutoff. Among the most efficient and popular is the injection of polymer-based gels in the formation of interest to reduce the water production and improve oil productivity. The success of such a program is highly dependent on the candidate selection quality. Ghoraishy et al. (2008) present a data-driven model able to predict the performance of future gel-treatment and select optimal candidates using historical treatment data. Two types of Bayesian networks (naïve and augmented) were trained and validated to predict the post-treatment performance from pretreatment data and the job design. A neural network approach (Saeedi et al. 2006) has also been proposed using only pretreatment well data as input to accurately predict post-treatment cumulative oil production with better accuracy levels than anecdotal screening guidelines.

The advantage of these systems is their ability to capture hidden information in the data and develop a model that, in the absence of any other numerical or analytical solution, can provide an excellent way to successfully screen and select candidates.

In waterflooding operations, with high-dimensional data consisting of many geologic and real-time operational attributes, conventional approaches such as simulation or deterministic analytical models can be cumbersome, time consuming, and not necessarily practical for tactical decision making. A data-driven modeling approach provides a practical alternative for quantitative ranking of different operating areas and assessment of production performance, development strategies, and optimization in heterogeneous reservoirs.

Real-time data from injector and producer wells, collected at the wellhead, pump-off-controllers, and fiber optics, are also used to understand vertical and areal conformance, to optimize injection profiles and rates at pattern levels (Maqui et al. 2017; Temizel et al. 2018). Pulse injection can also be used to understand interwell connectivity as an alternative to the well-established CRM when the latter cannot be applied because of lack of resources.

An intelligent production modeling workflow (Mohaghegh et al. 2014) using historical production and injection data and operational constraints has been used for maximizing production output from a field in the Middle East. The approach involves establishing the interwell connectivity using Voronoi mapping and models the expected production output and oil/water ratio by relaxing the choke opening. A neural network model is trained on field data and enables decision making on choke settings to maximize field production while minimizing the water output.

4.5.2.3. Capacitance/Resistance Model. The CRM, described earlier in this book, is a powerful yet simple semianalytical modeling approach for quick and robust evaluation of waterflooding performance. At its core, it is a generalized nonlinear multivariate regression technique that has its roots in classical signal processing. A change in rate at an injector creates a signal, which can be felt by one or more nearby producers (Sayarpour et al. 2007). In contrast to numerical simulators, which require a significant amount of geologic and operational data, CRM uses only production and injection data to history match and ultimately predict performance.

The data-driven component of the approach consists of exploiting the historical injection and production data to determine the model parameters that reflect the connectivity between the wells. Multiple solutions were developed and applied for practical purposes; however, the three most practical for reservoir control volumes are volume of the entire field, drainage volume of each producer, and drainage volume between each injector/producer pair (Sayarpour 2008). In these formulations, CRM can be applied for different scenarios such as full field analysis, a single well, or a group of wells. The capabilities of CRMs have been validated with both synthetic and real case studies. When calibrated with historical data, the model-generated solutions are comparable with those obtained from complex 3D numerical simulators. In the case of the real-field study, CRM was able to quickly history match field performance and it was used for prediction and optimization, thus leading to water reallocation and an 8% increase in production.

CRM has also been used to establish interwell connectivity from production- and injection-rate variations (Yousef et al. 2006). The methodology was applied for validation in two fields—onshore Argentina and Magnus Field in the North Sea. The approach indicated different flow characteristics that seem to agree with the presence of known geological features.

CRM has also been expanded to integrate with DCA to improve the understanding of the behavior of a mature naturally fractured carbonate reservoir under gas injection (Salazar-Bustamante et al. 2012). This approach addresses the short-coming of applying the CRM alone in the case of primary depletion followed by a weak injection. Therefore, the integration of the two allows the DCA to capture the contribution of the primary depletion, while the CRM seizes the injection component of the field performance. The advantage of this approach is that it is data driven, relying entirely on the production/injection history. The capability of this DCA-CRM model was demonstrated on a deep carbonate naturally fractured reservoir under hydrocarbon gas and nitrogen injection; high reliability for short-term production predictions was demonstrated, while allowing fast workflows and interpretations.

Improvements of the classical CRM have been proposed (Holanda et al. 2015), using a linear system of statespace (SS) equations to define the relationships between inputs, outputs, and states that describe the dynamics of the system. As such, the SS-CRM is a multi-input/multioutput matrix representation that provides more insights into reservoir behavior than analyzing only well-by-well performance. The authors introduce three CRM representations and contrast their performance—namely, integrated, producer based, and injector/producer based (CRMIP). The work demonstrated that the highest accuracy and performance was observed in the CRMIP, which was able to better capture the heterogeneous areas and channel-like deposits.

CRM models are also used as a predictive model for waterflood performance diagnostics and optimization (Kansao et al. 2017). In this work, the CRM was generated to develop a forecast by matching historical production and injection data, followed by uncertainty analysis and optimization of a heterogeneous reservoir undergoing a large waterflood development. The study demonstrated how the CRM was used to identify water injection changes that led to increased oil production, while maintaining or reducing the water cut.

CRM has also been compared with streamline-based methods that provide an effective means to assess flow patterns and well allocation factors (Ballin et al. 2012). That study concluded that neither method was sufficient by itself, and the best strategy was to integrate them. The estimates of allocation factors from CRM were influenced by data quality and quantity over the fit interval, while the streamline-based method had intrinsic model uncertainty.

4.5.3. Steamflood Management. Steamflooding, known as thermal EOR, is the process in which heat is injected into a hydrocarbon reservoir for reducing the viscosity of the fluids and therefore increasing the ultimate recovery. This practice is generally applied in heavy- and extraheavy-oil reservoirs that cannot be produced through an open mine.

Like the waterflooding process, the development of a heavy-oil reservoir can be designed in different ways depending on the properties of the reservoir and the fluids, and the overall economics of the project. There are three types of development currently applied:

- Continuous steam injection in dedicated vertical injectors known as conventional steamflooding
- Huff 'n' puff or cyclic steaming operations, where wells act as both injectors and producers in cycles

- Continuous steam injection in dedicated horizontal steam injectors drilled in pairs with dedicated horizontal producers placed just below them. This method is known as SAGD.

4.5.3.1. Steamflooding. Steamflooding operations present another fertile ground for data-driven analytics. While somewhat like waterflooding, given the continuous injection of agents (steam vs. water) through dedicated wells, it is significantly different from the standpoint of the recovery mechanism. Rather than displacing the oil, the steam tends to rise to the top of the reservoir forming a steam chest, allowing the oil to reduce in viscosity and drain into the producing wells. The production prediction of such steamflood developments is challenging for either analytical calculations or numerical simulators. Most of the development projects are therefore forecast using past analog developments. In large heavy-oil fields with multiple formations, data-driven models such as neural networks can be trained on the basis of past performance to predict the expected production outcome of a new development. These models not only can augment the analog P10/P50/P90 models and numerical reservoir simulations, but also serve as proxies for sensitivity analysis and grounding. Such models have been developed and successfully used in several fields in the San Joaquin Valley in California.

An additional application of data-driven models applies to steam-injection redistribution to optimize injection and maximize reservoir performance. It is well-understood and accepted that because of the inherited heterogeneities of the reservoir and the nature of the operation conditions (steam volumes, distribution, and qualities), the steam-chest development is not uniform. Therefore, multiobjective optimization algorithms and cloud computing can be used to run hundreds of thousands of scenarios to optimize the steam redistribution and maximize production (Ma et al. 2015; Sarma et al. 2017a, 2017b).

4.5.3.2. Cyclic Steam Operations. The nature of cyclic steam stimulations, or huff 'n' puff, presents a wealth of historical information in terms of steam injection volumes, rates, and pressure and corresponding production response. The use of cyclic steam operations dates to the early 1960s in the heavy-oil reservoirs of California. Numerous authors proposed different analytical models. Additionally, numerical simulators were introduced to better understand the production response and improve ultimate recovery. Regardless of the approach, both analytical and numerical models need to be calibrated using history matching, which presents a challenge given the dynamic nature of the cyclic process.

Data-driven technologies offer another approach to cyclic steam modeling and optimization, including candidate selection and steam volume optimization. A novel approach called data physics (Zhao and Sarma 2018) presents a framework to quantitatively optimize performance from a reservoir. The methodology integrates fuzzy logic, neural networks, and other machine learning techniques with the physics of reservoir simulation, therefore respecting mass balance, thermodynamics, and Darcy's law. A modified ensemble Kalman filter was implemented for data assimilation from thousands of wells and different data sources. Statistical techniques were used to demonstrate the predictive capability of the calibrated models, whereby the predicted performance of the steam jobs was compared to actual production observed in the field. The application of the workflow to a heavy-oil field in the San

Joaquin Valley demonstrated that the approach delivered an improved well selection for cyclic steam and an optimized injected steam volume. An incremental-production response of 44% was achieved, yielding a 77% profitability increase (Sarma et al. 2017a).

A different, yet related data-driven approach was used to mine the cyclic steam performance of the Cat Canyon Field. Historical steam injection volumes, rates, quality, injection duration, and corresponding production response were collected from more than 600 wells operated in the field. A data mining approach was used to discover patterns of injection/production performance. Using patterns information, a neural network was trained to predict the expected production outcome from the wells, thus delivering a ranking of the best candidates to be placed on steam on any given day. A better well ranking and selection was achieved, which was additionally combined with an optimized implementation schedule, which delivered significant incremental value to the field.

4.5.3.3. Steam-Assisted Gravity Drainage. The SAGD process consists of a pair of horizontal wells, with continuous steam injection being placed in a dedicated injector well and with the producer well placed just below it collecting the gross production (oil and water). Data-driven modeling has also been applied for recovery performance prediction in an SAGD application (Amirian et al. 2014; Dzurman et al. 2013). The objective of this work was to short-cut the detailed and time-consuming solutions provided by numerical simulators and replace them with a data-driven proxy model able to quickly screen different areas of the reservoir, assess the uncertainty resulting from heterogeneity, and provide a quantitative ranking for the best development. Thus, an ANN was trained on the basis of a training data set generated by running a numerical flow simulator in various scenarios. The scenarios were run considering a wide range of values for the attributes describing the characteristics associated with the reservoir heterogeneities and relevant injection/production design. While theoretical in nature, the approach is interesting and can be easily adopted by companies that operate many SAGD pairs across larger field areas.

The advancement in technology for surveillance, such as fiber optics, brought a completely new and exciting opportunity for data-driven applications for SAGD operations. The distributed acoustic sensing generates millions of data points for attributes such as temperature and pressure along horizontal wells at frequencies as high as a couple of seconds. This information, when mined properly, can provide a superior understanding of the steam distribution along the horizontal, the extent of the steam chambers, and ultimately the production performance. Algorithms, such as deep learning, can process these streams of data and assist not only in well workovers to optimize steam distribution and heating, but also in early reservoir diagnostics.

Summary

- *Data analytics has been used in the context of analog selection to provide more-robust predictions of the expected reservoir performance.*
- *Various data analytics have been developed to optimize various aspects of mature waterfloods such as new well targets, water shutoff, or continuous optimization target rates for production and injection wells.*

- *Steamflooding has also benefited from analytics where neural networks have been trained on historical data to help forecast the performance of new projects. Another approach consists of training neural networks on the basis of simulation results to provide a fast and accurate prediction engine for steamflood optimization.*

4.6. Reservoir Simulation. Reservoir simulation is one of the most popular methods for making predictions of reservoir performance under different development and operation strategies. The results of performance prediction serve as the basis for many capital-intensive investment or reservoir management decisions. Reservoir simulation is often computationally intensive because of the large number of cells required to represent the whole reservoir and/ or the complex physics such as multiple phase/multiple component and coupling with the surface network or with geomechanics. It is common that a full reservoir simulation run takes many hours to days even with most advanced parallel solution using multiple cores in a high-performance computing (HPC) environment.

On the other hand, the inputs of reservoir simulation including reservoir static and dynamic properties are mostly uncertain because of limited sampling or error in the measurement. This results in uncertainty in performance predictions. It has been well-recognized that important reservoir management decisions need to account for these uncertainties to better manage the risk. Quantifying uncertainties in performance prediction often requires performing a large number (hundreds to thousands) of simulations, which adds additional burden on reservoir simulation. Similarly, model calibration (history matching) and reservoir development optimization also often require many simulation runs by using different combinations of model-parameter values and development/well operation strategies.

Data analytics has been widely used to reduce such burdens in reducing the computation time and costs. The two main areas that are widely used or researched on are proxy modeling and reduced-order modeling.

This section discusses the data analytics methods applied to model generated data including proxy modeling, reduced-order modeling, and ensemble variance analysis, followed by introducing the latest attempts in the development of data-driven physics-constrained predictive modeling.

4.6.1. Proxy Modeling. Proxy modeling (or the response-surface method) explores the relationships between several explanatory variables and one or more response variables. This section discusses the various proxy modeling methods and best practices in practical applications.

4.6.1.1. Proxy Model Types. Proxy models are designed to approximate the response of a system by a simple mathematical function of its input parameters without running the computationally intensive simulations. The advantages of having a proxy to approximate a simulation include using a proxy to instantly predict the model behavior at unsampled locations, and using a proxy to quantify the impact of a change in input parameters on the output. Proxies work best when the system response is smooth. They should be tested for validity using cross validation.

Proxies are usually built on the basis of a small number of actual simulation runs (samples). A popular method to select the samples is experimental design (ED) or design of experiments (DoE) where the minimum number of simulation runs are selected to obtain maximum information based on the uncertainty space (Montgomery 2012). For more details about ED or DoE in petroleum engineering applications, readers can refer to Friedmann et al. (2003) or Yeten et al. (2005).

Several types of proxies are used in practice. Some are based on simple analytical functions, while others use numerical approximation that cannot necessarily be represented explicitly by a simple function.

An analytical proxy approximates the relationship between input factors and output response by an analytical function, such as a polynomial. The proxy is constructed by fitting the function to the data points using regression techniques (i.e., least squares). It determines the coefficients of the analytical function by minimizing the sum of the squares of the errors between the data points and the fitted function values. Each data point represents a simulation run selected from the design matrix. If the number of data points equals the number of unknown coefficients of the analytical function, the proxy will traverse all the data points. If this is the case, the proxy is data exact. However, because the proxy is only an approximation, more data points than the number of coefficients should be used to reduce error. In other words, the least-squares method should be applied to solve overdetermined problems. Hence, analytical proxies are often not data exact.

A numerical proxy approximates the relationship between input factors and output response by attempting to connect all the data points using a surface that is generated from a numerical algorithm and cannot be represented by an analytical function. Such a proxy is called data exact if it traverses all the data points. Three commonly used numerical proxies are Kriging, splines, and neural networks.

Kriging predicts the value of a function at a sampling point by calculating the weighted average of the function values of the data points. Kriging assumes that all the points are spatially correlated with each other. The correlation is described by a variogram model, (h), which is a function of the Euclidean distance (h) between two points. The larger the distance, the more variant the two points are. The weights for computing the weighted average are obtained from the covariance between the sampling point and each data point and between all the data points themselves. A larger weight is assigned to a data point if it is closer to the sampling point. The weights are computed such that the squared error of the estimated function value of the sampling point is the smallest. Kriging is data exact. More details can be found in any geostatistics textbook, such as Journel and Huijbregts (1978) or Deutsch and Journel (1992).

A spline function is defined by piecewise polynomials and has a high level of smoothness at the knots where polynomials connect. Each knot refers to a data point whose function value is known. More details on spline-based proxies can be found in Li and Friedmann (2005).

ANNs are machine learning methods that mimic the operations of the cortical systems of animals. An ANN model consists of many interconnected nodes, like the neurons inside a brain. Each node accepts inputs from other nodes and generates outputs that are based on the inputs it receives and/or the information stored internally. These nodes are grouped in multiple layers: an input and an output layer, and one or more hidden layers in the middle. The input layer is analogous to a sense organ, such as eye, ear, or nose, which receives inputs from outside. The hidden layers process the inputs to produce the corresponding response that is then reported by the output layer. Like a human being, an ANN needs to be trained to produce proper output response on the basis of given inputs. The training is done using the data points where the relationship between input factors and output responses is known. ANNs are often not data exact. More theoretical and algorithmic details can be found in Reed and Marks (1999).

Other popular machine learning methods include random forest, support vector machine, or gradient boost methods. Details of these machine learning algorithms can be found in Mishra and Datta-Gupta (2017).

4.6.1.2. Best Practices for Proxy Modeling. Extrapolation. In theory, proxies only used for interpolation and extrapolation should be avoided. It is, however, possible to extrapolate without realizing it, especially in high-dimensional problems where a good coverage of the search space requires many experiments. As an example, in **Fig. 4.13** a linear proxy was generated from the three points marked by black dots. The proxy is used in interpolation over the green area but in extrapolation over the red area. This is sometimes hard to detect, because from a univariate standpoint, the points located in the red area appear to be within the range of data used to define the proxy. Detecting these effects in a high-dimensional problem is a challenge. These extrapolations are often performed during Monte Carlo sampling and should be performed with caution.

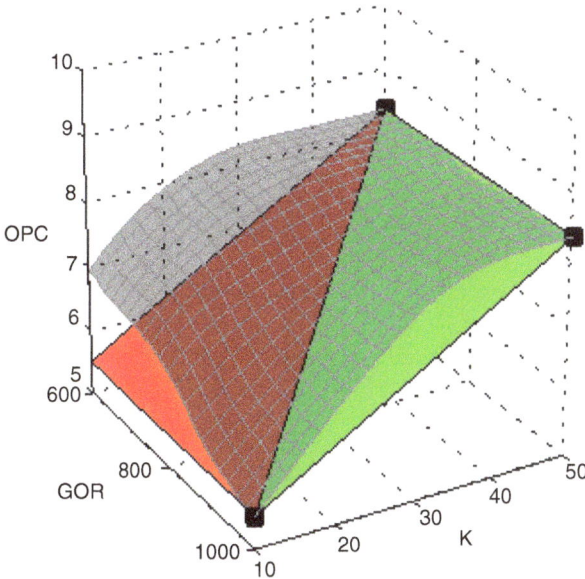

Fig. 4.13—Linear proxy model.

Cross Validation. One of the most important practices when building proxies is to implement a verification process using blind tests. Blind tests are sample points where a simulation model is run, but the response is not used in the creation of the proxy. Instead, the proxy is used to estimate the response at the sample location, which is compared to the calculated value. Blind tests are critical to validate a proxy but are sometimes deceiving because a successful blind test only guarantees a robust proxy in a local part of the domain. Some common methods for cross validation are

- Exclusion of some experiments during the calibration of the training, so that the experiments can later be used as blind tests. Such practice should be used with designs that offer some redundancy because it might deteriorate the sampling.
- Addition of some experiments that can be chosen using ED or Monte Carlo sampling.
- Leave-one-out cross validation, where every run is used as a test point for the others. For each run performed, a proxy is built using the other runs of the design. The proxy is then used to estimate the response, which is compared to the actual response from the run. The algorithm loops through the runs, so that all runs are used alternatively as test and calibration points. This method has been extended to consider more than one experiment at a time, which is known as k-fold cross validation.

Fig. 4.14 represents a graphical example of a cross validation. The ED points are shown in black, and the validation points are shown in red.

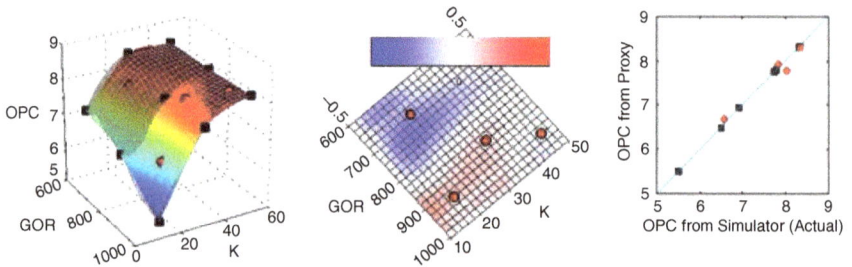

Fig. 4.14—Cross validation.

Overfitting. It is important to match the level of complexity of a proxy to the number of experiments used. Using a simplistic proxy fails to properly represent the model response, but using a proxy that is too complex can lead to overfitting and can generate inaccurate behavior. When multiple proxy models are built, the simpler models passing the validation tests should be preferred.

Minimize the Dimensionality. The number of experiments needed for a proxy grows exponentially with the dimensionality of the parameter space. It is best to keep the parameter spaces as small as possible to build robust proxies and keep the number of experiments required to a minimum.

Use Multiple Proxies at the Same Time. It has been recommended to build several proxies together (Schuetter et al. 2015). Combining proxies built with neural networks, Kriging, and splines can provide more-robust estimations than any single proxy.

Applications of proxy can be widely found in reservoir engineering for uncertainty quantification, history matching, and optimizations (Friedmann et al. 2003; Landa and Güyagüler 2003; Bhark and Dehghani 2014; Yeten et al. 2005; Amudo et al. 2014; Sarma et al. 2015; He et al. 2015a, 2015b, 2016a, 2016b; Chen et al. 2017; Wantawin et al. 2017; Pankaj et al. 2018).

Friedmann et al. (2003) applied the response-surface methodology for evaluation of channelized sandstone reservoirs under input uncertainty. The method proposes a two-step process, starting with Plackett-Burman analysis to determine significant parameters and response curvature, followed by D-optimal analysis to identify quadratic nonlinearities and interactions. The polynomial models generated using ED methodology (**Fig. 4.15**) serve as good analogs to reservoir simulators to evaluate production-forecast uncertainty.

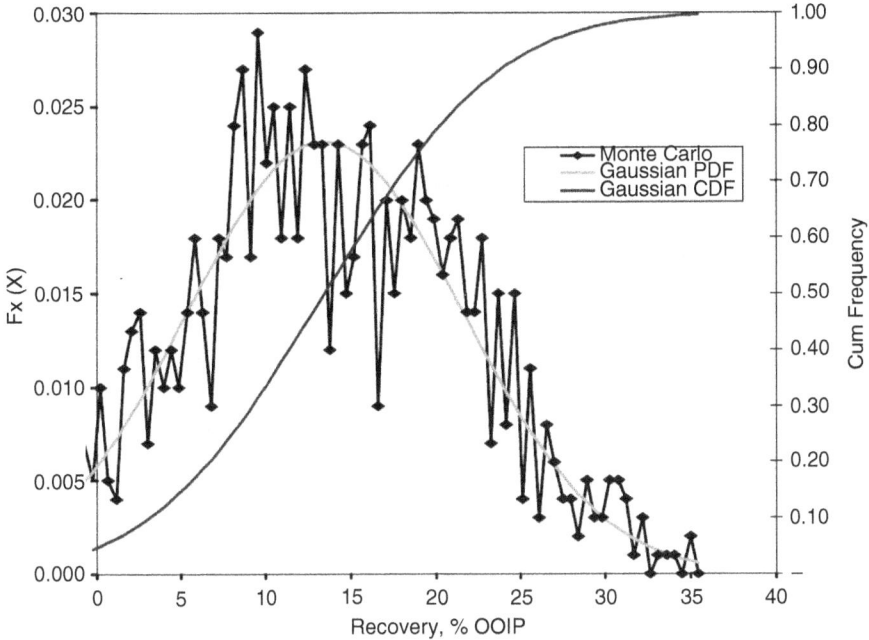

Fig. 4.15—Probability density function (PDF) and cumulative distribution function (CDF) for primary depletion of channelized reservoirs undergoing bottomdrive (Friedmann et al. 2003)

Iterative Response Surface. The common process of proxy building based on a set of sampling might not be sufficient to build reliable and accurate proxies, particularly when the problem is nonlinear. One way to improve this is through an iterative process where the new sampling points are iteratively added to update proxies. Some optimization algorithm is used to optimize the locations of the most informative points between iterations (Castellini et al. 2010; Wantawin et al. 2017).

To initialize the algorithm, any appropriate ED can be used as the initial sampling points, such as Plackett- Burman, D-optimal, Latin hypercube, or Hamersley sequence. After the initial proxy is built, the following proxy properties can be considered when selecting the new sampling points: function value, scalar gradient, bending energy, curvature, and distance from existing points. A combined score can be computed for each proposed new sampling, and the best samplings can be selected. The procedure stops when the number of iterations requested by the user is complete or when the changes to the response surface are below a certain threshold.

For example, **Fig. 4.16** shows three iterations of the proxy building process. A spline-based proxy is initially built using nine samples (shown as the red points

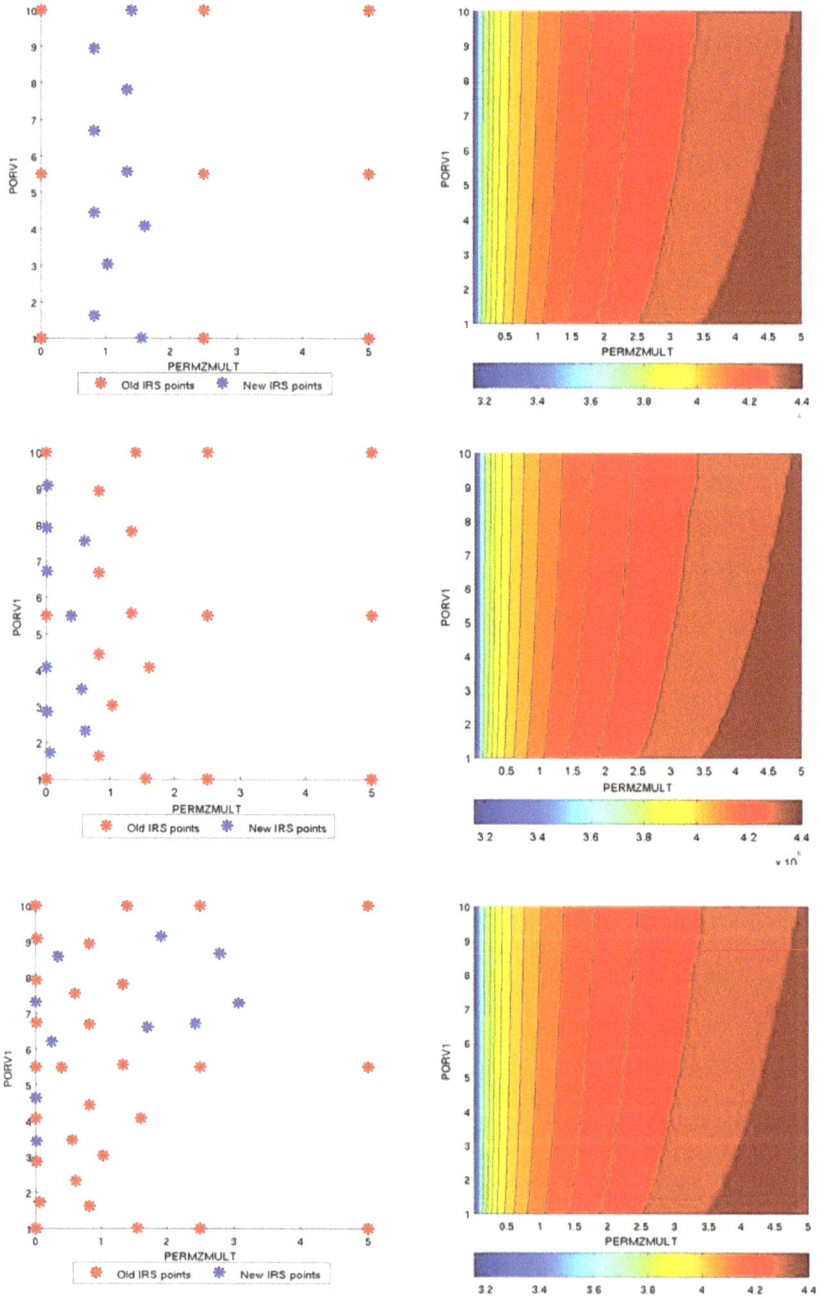

Fig. 4.16—Iterative proxy generation; first three iterations (rows) showing the sample locations (red for current and blue for next sample locations) and proxy shape (right).

in the top row). On the basis of the initial proxy, an additional 10 samples (blue points) are selected that are based on multiple selection criteria. New simulations are performed, and the proxy is updated with the new samples. This process is repeated until a stable proxy is constructed or a maximum number of iterations is reached.

4.6.1.3. Probabilistic Collocation Method. Proxies built using DoE do not account for the distributions of the parameters. The probabilistic collocation method defines optimal sampling points that are based on the probability distribution function (PDF) of the parameters using polynomial chaos expansion (PCE). The proxy is then built using the PCE terms derived from the distribution functions, with the coefficients being estimated through regression with the samples (Gautschi 1994; Tatang et al. 1997; Xiu and Karniadakis 2002; Li and Zhang 2007, 2009; Sarma and Xie 2011).

The PCM is a technique that uses the PDFs of the input factors to guide ED and proxy construction in one step. Its greatest advantage is that a reasonably accurate proxy can be built in much fewer runs than are required for a traditional modeling design.

PCM can be considered as an advanced ED technique with proxy construction embedded in it. By taking the input-factor PDFs into account, the final proxy is more accurate in high-probability regions than the traditional EDs are. Studies have shown that the resulting proxies are more accurate and more stable compared with those built by ED sampling, particularly when the problem is nonlinear (Li and Friedmann 2005, Li et al. 2011).

Another advanced method is nonintrusive spectral projection, which can be used for further improvement of proxy modeling (Le Maitre et al. 2001; Reagan et al. 2004; Sarma and Xie 2011).

4.6.2. Reduced-Order Modeling. Like the proxy model described in the preceding section, reduced-order modeling can be considered as other surrogate models can be implemented in place of the traditional reservoir simulator for computationally intensive applications such as production optimization and history matching. Reduced-order models apply fast but approximate numerical solutions that are consistent with the underlying governing equations, which is the main difference from the proxy modeling described in the preceding section.

Reduced-order modeling procedures, which have been applied in many application areas including reservoir simulation, represent a promising means for constructing efficient surrogate models. Many of these techniques entail the projection of the full-order (high-fidelity) numerical description onto a low-dimensional subspace, which reduces the number of unknowns that must be computed at each timestep. We can classify existing approaches applied within the context of reservoir simulation as grid-based methods, system-based methods, and snapshot-based methods (He 2013).

In grid-based methods, constructing a coarser grid and then computing properties for this grid reduce the dimension of the problem. The original problem is then solved on the coarser grid. Examples are upscaling and multiscale methods. System-based methods are derived from system control theory. By introducing a basis matrix and a constraint reduction matrix, both generated from the full-dimensional system, a full-dimensional state matrix can be reduced to a much-lower-dimensional linear time-invariant system, which can be solved much more efficiently (Bond and Daniel 2008).

We will focus on the description of snapshot-based methods that are more commonly used in the reservoir simulation community. Unlike system-based methods, which derive the basis and reduction matrices from the system matrices, snapshot-based methods derive the basis matrices from snapshots, which are the states at each timestep of training simulations. Most methods in this category are based on proper orthogonal decomposition (POD) or its variants.

4.6.2.1. Proper Orthogonal Decomposition. POD reduces the dimension of the problem by projecting the high-dimensional states (e.g., pressure and saturation at each gridblock in an oil/water problem) onto an optimal lower-dimensional subspace. The basis of this subspace is obtained by performing singular value decomposition (e.g., Golub and Van Loan 1996) of a matrix containing, as its columns, the solution states (snapshots) computed from the previous simulations. The simulations used to provide these snapshots are the training runs; new (prediction) runs are referred to as test cases.

Van Doren et al. (2006) applied POD to reduce the dimensions of both the forward model and the adjoint model with the goal of accelerating the optimization of a waterflood process. A 35% reduction in computing time was reported in that work. Cardoso et al. (2009) proposed a snapshot clustering procedure and a missing point estimation technique to further accelerate a POD-based reduced-order reservoir simulation model. They achieved speedups by a factor of 6 to 10. The speedups achieved in these studies are quite modest because the POD-based methods required the construction and projection of the full Jacobian matrix at each iteration of the simulation. The computational complexity of these operations is on the order of that of the original problem, which limits the amount of speedup that can be achieved. Alternative treatments of nonlinearity are needed to achieve larger speedups. In addition, unlike system-based methods that calculate both basis and reduced matrices, POD provides only the basis matrix, which reduces the dimension of the states. The appropriate constraint reduction matrix, which reduces the number of equations, still must be determined. This led to the development of trajectory piecewise linearization (TPWL).

4.6.2.2. Trajectory Piecewise Linearization. TPWL is introduced for efficient treatment of nonlinearities and thus for achieving larger speedups. With this approach the solution at each timestep of the test simulation is represented in terms of a linearization around the closest saved state computed during the training simulation. TPWL was first introduced by Rewienski and White (2003). It was combined with a Krylov subspace order-reduction approach and applied for circuit design problems. Cardoso and Durlofsky (2010) first considered the use of TPWL for oil/water reservoir simulation problems. They used POD for order reduction rather than a Krylov method, which gave rise to a POD-TPWL procedure. As noted earlier, the basis from POD is constructed from the states computed during training simulations, and it generally does not require updating. In addition, the optimality of POD allows for a high degree of reduction and high speedups. The results in Cardoso and Durlofsky (2010) showed reasonable solution accuracy and substantial speedups, on the order of a factor of 500, for a variety of test cases. The use of TPWL for optimization of well controls (time-varying bottomhole pressures) was also presented. In subsequent work by He et al. (2013) and He (2010), the accuracy and stability of the POD-TPWL method were further enhanced. In that work, local resolution and basis optimization procedures were introduced, and the resulting POD-TPWL procedure was successfully applied to oil/water models that contained more than 100,000 gridblocks. POD-TPWL approaches have also been developed for idealized thermal models (Rewienski and White 2003; Rousset et al. 2014). These methods have not, however, been applied for compositional problems.

When dimension reduction is applied using a projection matrix, the number of state variables in the system is reduced. This results in an overdetermined system with more equations than unknown variables. This overdetermined system is rendered solvable by premultiplying the system by the constraint reduction matrix that defines the subspace in which the residual of the system is driven to zero.

In summary, POD-TPWL combines TPWL and POD to provide highly efficient surrogate models (**Fig. 4.17**). The POD-TPWL method expresses new solutions in terms of linearization around states generated (and saved) during previously simulated training runs. High-dimensional states (e.g., pressure and saturation in every gridblock in an oil/water problem) are projected optimally onto a low-dimensional subspace using POD (He and Durlofsky 2010; He et al. 2011a, 2011b; He et al. 2013).

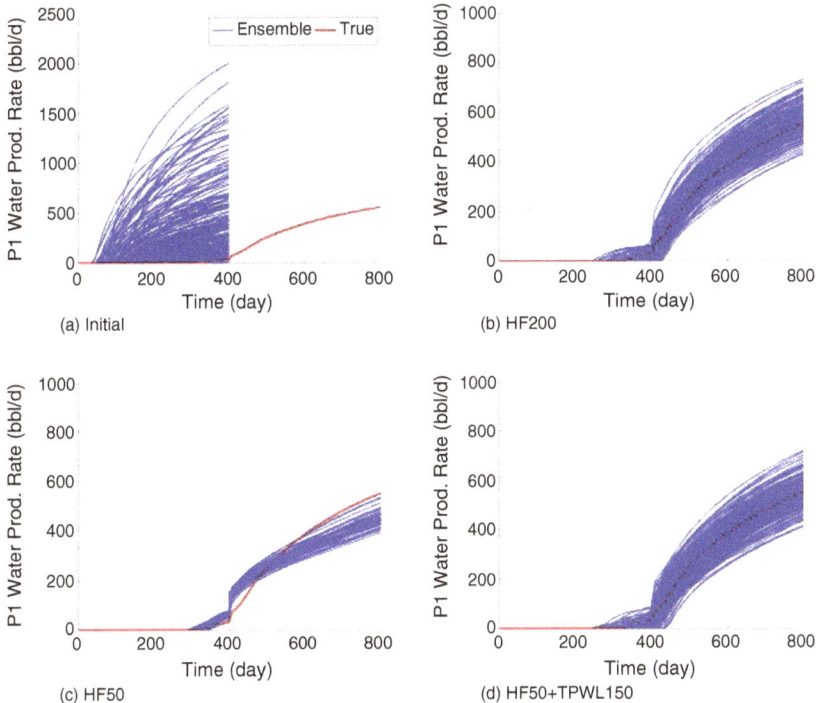

Fig. 4.17—Production profiles and history match for a producer water rate using high fidelity (HF) ensembles and TPWL (He et al. 2011a).

4.6.2.3. Discrete Empirical Interpolation Method. Though POD is frequently applied to nonlinear problems, it should be noted that this only optimally approximates linear manifold in the configuration space represented by the data. From a practical standpoint, the dimension reduction of nonlinear terms involves a complexity proportional to the number of reduced parameters. In this regard, the discrete empirical interpolation method (DEIM) provides a dimension reduction of the nonlinear terms that has a complexity proportional to the number of reduced variables (Chaturantabat and Sorensen 2009).

POD-DEIM methods combined with radial basis function were used in a nonintrusive manner to describe the reservoir dynamics entailed by multiple combinations of inputs and controls (Klie 2013). A global/local model reduction with POD-DEIM was introduced with an auxiliary variable (velocity field) for high compression of the model (Ghasemi et al. 2015; Yang et al. 2016). A multiscale method allows POD snapshots to be inexpensively computed with a hierarchical approximation of snapshot vectors and adaptive computations on coarse grids. Further, this method was extended to develop an online adaptive POD-DEIM model reduction by incorporating new data as they become available (Yang et al. 2017). Another variant using trajectory-based DEIM was used to approximate the nonlinear terms in the test simulation as a sum of nonlinear terms evaluated at the closest available training points from the high-fidelity training trajectory and a perturbed term defined as the difference between the test and the training terms (Tan et al. 2017).

4.6.3. Reduced-Physics Model. Another proxy model that we will not discuss in this book is reduced-physics models that accelerate flow simulations by simplifying the physics. Streamline methods (Batycky et al. 1997; Datta-Gupta and King 2007) fall into this category. Streamline methods decouple the flow and transport equations and then solve the transport equations as a series of 1D problems along each streamline. This simplification can lead to substantial speedups relative to traditional simulation for some problems.

Streamline methods have been applied for a wide range of problems including production optimization (Samier et al. 2002; Thiele and Batycky 2003; Tanaka et al. 2017) and history matching (Milliken et al. 2001; Wen et al. 2003). These approaches approximate many key effects, and though they have been widely used for waterflooding applications, they are not commonly applied for compositional problems. In addition, the overall speedup using streamline methods is still limited because of the need to solve the full-order equations at some timesteps.

More recently, the concept of DTOF has been extended to calculate the propagation for the pressure front in the reservoir for black-oil (Xie et al. 2012; Zhang et al. 2016; Lino et al. 2017) and compositional simulation (Lino et al. 2017). The approach consists of two decoupled steps—calculation of the DTOF using the fast-marching method and fully implicit simulation using DTOF as a spatial coordinate (Lino et al. 2017a, 2017b). The computational efficiency is achieved by reducing the 3D flow equations into 1D equations using the DTOF as spatial coordinate, leading to orders of magnitude faster computation over full 3D simulation. Computational-time savings also increase significantly with grid refinement and for high-resolution models.

4.6.4. Predictive Uncertainty Analysis. A more recent advance in data analytics applied to model-generated data is the so-called predictive uncertainty analysis method, designed to support rapid decision making under uncertainty. Typical use cases are

- Design optimal pilot program or surveillance plan to maximize value of information before the actual data are collected.
- Quickly update performance predictions without needing to follow the traditional model calibration process to accelerate subsequent reservoir management decisions. In this approach, a series of simulation runs are performed using some sampling strategy in the parameter and/or operating space (such as DoE/ED) and the input and response parameters are recorded in a database.

The development of oil and gas reservoirs is associated with substantial risk because the subsurface condition is highly uncertain. Data acquisition programs such as surveillance and pilots are routinely conducted in the hope of minimizing subsurface uncertainties and improving decision quality. However, these programs themselves involve significant capital investment. Therefore, before any data acquisition program is implemented, it is crucial to be able to evaluate the effectiveness and quantify the value of the program so that different designs can be compared, and the best investment decision can be made. The major challenge of estimating the effectiveness of data acquisition programs is that the data are unknown at the time of the analysis.

As surveillance data are obtained from the field, the cumulative probability distributions (CDFs) of the key metrics need to be updated accordingly. This is normally accomplished by a two-step approach as shown on the left side of **Fig. 4.18**. First, the data are assimilated through history matching to calibrate the model parameter uncertainties to obtain their posterior distributions. Then, a probabilistic forecast is performed on the basis of the posterior distributions of the parameters to update the S-curve of the key metrics.

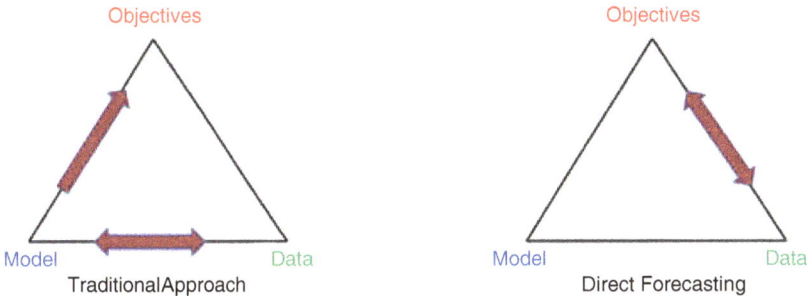

Fig. 4.18—Traditional model-driven approach (left) vs. direct forecasting approach (right) for prediction and update of prediction using observation data.

This two-step approach can be very time-consuming. This is because the relations between the objectives and the model parameters, and the relations between the data and the model parameters, are often highly nonlinear. In addition, the potentially large number of model parameters makes it very hard for any history-matching algorithms to accurately capture the posterior distribution of the model with a small number of simulation runs. Because of these challenges, updating the CDF with new data using the traditional approach can take weeks or months.

On the other hand, many of the field development decisions are time critical and there might not be enough time/resources to calibrate the model or to perform any simulations after data come in. In those cases, there is a need for rapid interpretation of the incoming data and updating of the S-curve without going through a full-blown history-matching and probabilistic-forecast process.

In a recent study, the approach called direct forecast or predictive uncertainty analysis (also called data-space inversion) (Scheidt et al. 2015; Satija and Caers 2015; Sun et al. 2017; He et al. 2015a, 2015b, 2017a, 2017b, 2018) has been proposed. Fig. 4.18 shows the concept of the direct forecast method on the right, which can be considered as a data analytics or machine learning approach using model-generated data.

In direct forecast, the statistical relationship between the proposed measurement data and the business objective is established based on simulation model responses before the data acquisition. This direct relationship can then be used to rapidly update the prediction of the objective once the data become available. These processes are illustrated in **Fig. 4.19**. Data analytics algorithms can be used to address these challenges (He et al. 2018).

4.6.5. Data-Driven Physics-Based Predictive Modeling. In this section, we discuss recent contributions to data-driven physics-based modeling, in which models based on simplified physics are built directly to fit the observed data. The purpose of data-driven physics-based modeling is to use a variety of simplified physics models that can explain the observed data. This is usually associated with a fast data assimilation method to quickly account for observation data by adjusting model parameters, such as the ensemble-based data assimilation method (Evensen 2004; Wen and Chen 2005, 2007; Aanonsen et al. 2009; Emerick and Reynolds 2013).

Historically, engineers model and predict reservoir performance using tools that vary in complexity in terms of modeling the underlying physics. These tools range from simple DCA to full-fledged 3D compositional reservoir simulation. For a reservoir engineering problem at hand, engineers must select appropriate tools according to both the nature of the problem and the practical logistics of the project. Quite often the best tool to apply is not the most-advanced or -sophisticated tool at hand.

In the traditional model-driven approach (left figure in **Fig. 4.20**), reservoir static models are often constructed by integrating a variety of data sources from geophysics, petrophysics, and geology. These full 3D static reservoir models are usually complex and do not include dynamic data such as those observed from wells (e.g., rate and pressure data). Reservoir performance is then performed through flow simulation based on the static model by assigning different development strategies to obtain the production profile that will be used for evaluating the economics. When there are observed data from production, these models will require calibration to confirm the observation data by means of a tedious history-matching process. The updated performance prediction is then obtained by running flow simulation of history-matched models.

Big data and advanced data analytics have opened another window for predictive modeling of reservoir performance using purely data-driven methods without modeling the underlying physics. These purely data-driven methods have been discussed

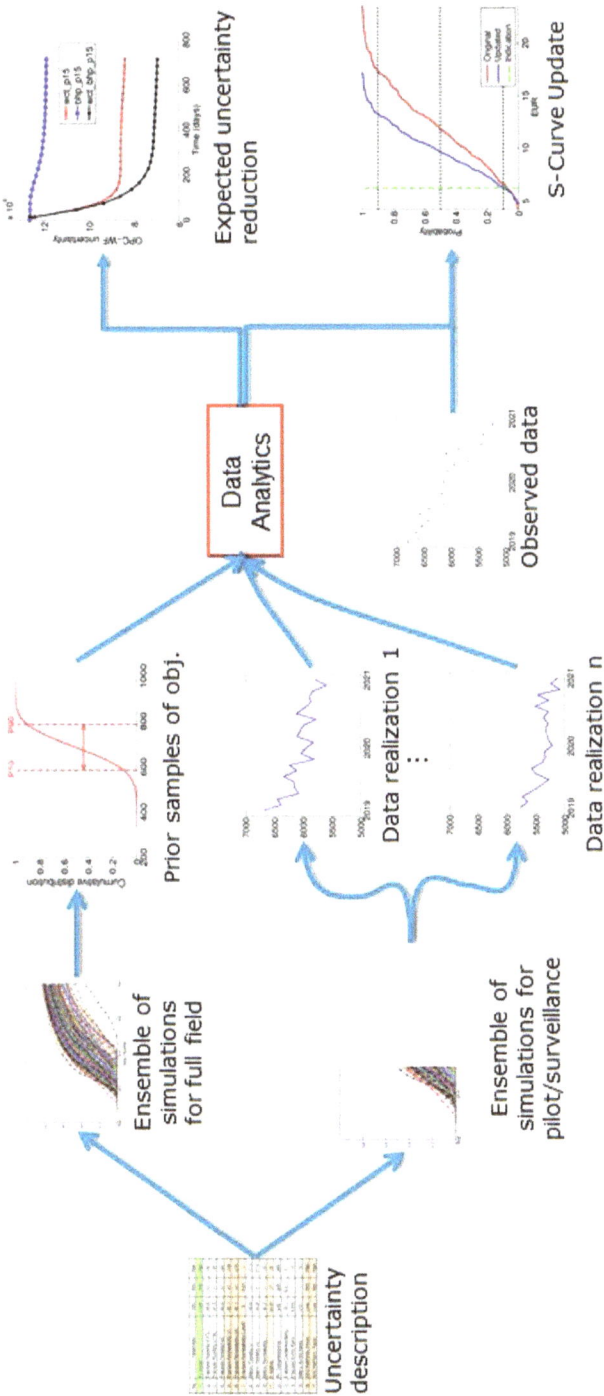

Fig. 4.19—Concept of applying data analytics to model-generated data to find leading indicator through uncertainty reduction (before data come in) and rapid updating of forecast (after data come in).

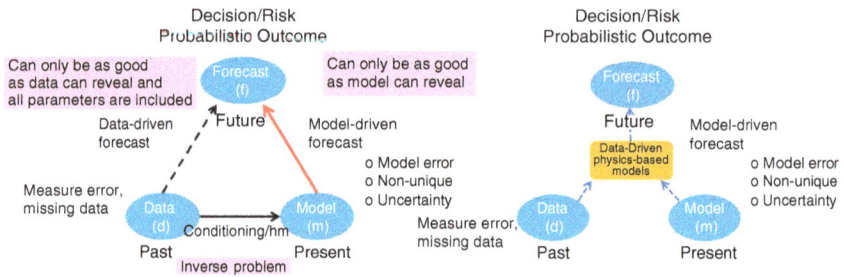

Fig. 4.20—Traditional model-driven physics-based approach (left) and data-driven physics-based approach (right).

in other sections (Section 3 and Section 5.6.1) of this book. Here, we focus on the approaches to combine conventional physics-based reservoir engineering models with big-data and data-driven technologies (right side of Fig. 4.20). In this approach, an ensemble of models (simple but sufficient to account for the important physics) are constructed on the basis of available data and known physics. Forecast is obtained directly after models are calibrated to the observed data by means of a simple and fast classification or data assimilation method. The predictive power of purely data-driven methods is often questionable outside the space covered by actual data. It is hopeful that with this physics-constrained hybrid approach, we can simultaneously obtain the high performance of the data-driven methods and the better predictability of physics-based models outside the data range. In the following subsections, we describe several emerging technologies that aim to achieve this goal.

4.6.5.1. Diffusive Diagnostic Function. Recently, Zhang et al. (2018) developed a physics-based data-driven model for history matching, prediction, and characterization of unconventional reservoirs. It uses 1D numerical simulation to approximate 3D problems. The 1D simulation is formulated in a dimensionless space by introducing a new diffusive diagnostic function (DDF). The DDF is physically related to the reservoir properties and the hydraulic-fracture geometries.

As illustrated in **Fig. 4.21,** the idea is to discretize the reservoir according to a set of contour surfaces of the unknown 3D solution. As a result, the reservoir is conceptually modeled as 1D. The key point here is that, for a given reservoir, we do not need to derive the 1D model by building a detailed 3D model. Instead, we just need to find possible 1D models that will reproduce the observed production history. Each possible 1D model is fully described by a DDF. The DDF has certain characteristics that are closely related to reservoir and completion parameters such as total fracture area and stimulated reservoir volume. This allows initial estimation of DDFs from prior knowledge of the reservoir and hydraulic fractures. It also allows post-history-matching diagnostics of these physical parameters. Because the forward simulation using the 1D model is extremely fast, thousands of realizations can be generated to achieve history matching in minutes. Compared with conventional reduced-physics methods such as DCA and RTA, the DDF approach can easily incorporate complex flow physics such as multiphase flow and a compositional fluid model without assuming any specific flow regime.

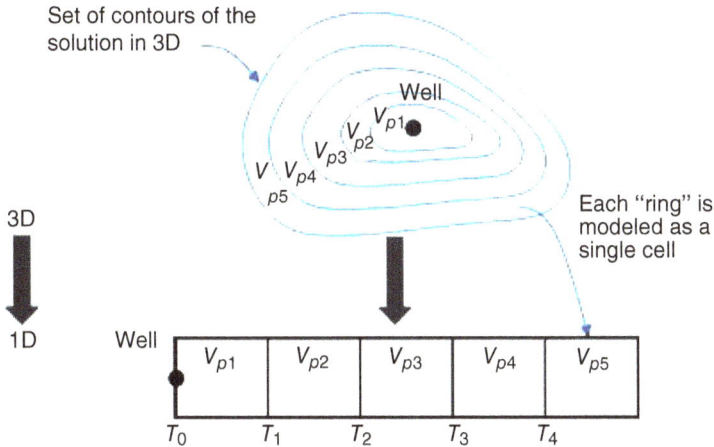

Fig. 4.21—Schematic to show how a 3D reservoir simulation problem could be reduced to a 1D problem.

4.6.5.2. Other Approaches. Other similar approaches include CRM (refer to Section 5.3.2.3) and INSIM-FT (Guo et al. 2017, 2018; Zhao et al. 2015, 2016a, 2016b), StellNet (Lutidze and Younis 2018; Ren et al. 2019), and the data/physics model (Sarma et al. 2017a, 2017b; Zhao and Sarma 2018). All these approaches try to simplify the reservoir model to connections between wells with different types of parameterization. Reservoir simulation or a semianalytical method can be used to account for flow physics for interwell modeling with a relatively small number of parameters, which results in fast turnaround. They all require significant historical data and rapid data assimilation methods or machine learning techniques to calibrate the model parameters before becoming a useful tool for prediction and optimization.

INSIM-FT. The reservoir is characterized as a coarse model consisting of several interwell control units where each unit connects a pair of wells. Each unit has two specific parameters: transmissibility and control pore volume. By solving the mass balance equations, for the control units, the interwell fluid rates and saturations are obtained so that phase producing rates can be predicted. A history-matched INSIM-FT model could reliably predict future production on a well-by-well basis accurately for both synthetic cases and a cross-validated field case. It has also shown that a history-matched INSIM-FT model could be used to estimate the well controls that maximize the NPV of production over any prescribed future time, generally the remaining life of the reservoir.

Other than its restriction to waterflooding applications, the only apparent defect of INSIM-FT is its restriction to 2D problems where all wells are fully penetrating vertical wells. It was then extended to 3D problems, referred to as INSIM-FT-3D.

The INSIM-FT-3D formulation was validated on synthetic examples, where a reservoir simulation model and a field case represent the reservoir. For these examples, INSIM-FT-3D was able to obtain good history matches and predictions.

StellNet. StellNet focuses on extending the idea of placing a 3D geocellular grid into a network of connections in an analogy of more-complicated physical scenarios, such as horizontal and multilateral well operations and steam- or gasflooding processes (Lutidze and Younis 2018; Ren et al. 2019). With the parallel implementation capabilities of the model and its generality to model physics, StellNet can replace a reservoir simulator for closed-loop optimization applications.

StellNet is a hybrid finite-volume and flow-network reservoir model including production facilities. The reservoir is characterized as a series of end nodes and connections between nodes. Here is how it works:

1. Generate a sparse, irregular 3D lattice of nodes to capture the geometry of structures such as completed well trajectories and fractures. To generate a graph, there are two types of points: well points, which are the nodes at the perforated zone along the well, and reservoir points, which represent the reservoir boundaries, aquifers, or other special features. A minimal undirected graph is then created, linking nodes to form a network.

2. Coupled with a wellbore model, the sequence of connections performs as a proxy of the reservoir. Each link is treated as a coarse, 1D simulation problem and solved with a finite-volume scheme. The connections can be divided into three major types: intrawell connection depicted by a wellbore model, interwell connection characterized by porous media flow between well nodes, and well/reservoir connection that captures the influence of reservoir boundaries and/or aquifers.

Data/Physics Model. Sarma et al. (2017) described a data/physics model that merges modern data science and the physics of reservoir simulation and allows quick setup and runs in real time. It focuses on predicting the production impact of certain activities, but without trying to characterize the details of reservoirs. The underlying reservoir physics equations (the same as in a reservoir simulator) are solved, which allows it to account for different physics including mass balance, heat balance, and Darcy's law. The coefficients of these equations in the model are calculated by means of a neural network. Production data, log data, and seismic data can be integrated through a single assimilation step using a modified ensemble Kalman filter (Evensen 2004; Wen and Chen 2005, 2007; Aanonsen et al. 2009; Emerick and Reynolds 2013).

Summary

- *Data analytics can be used in conjunction with numerical simulation to reduce computation time and cost.*
- *Proxy modeling, reduced-order modeling, reduced-physics modeling, and data/physics modeling are some of the modern data analytic approaches used with reservoir simulation.*

4.7. Unconventionals. Conventional reservoir engineering studies have typically been conducted with a combination of methodologies, including analogs, data-driven studies, analytical models, and numerical models. In conventional reservoirs, the physical laws governing the flow and transport problems through porous media are relatively well-understood and -documented. The modeling approach taken for a study is thus chosen on the basis of the data, time, and resources available.

The physics at play in unconventional reservoirs remains an area of active research. The complexity of the physical phenomena involved, combined with the scale and pace of development, has pushed the industry to adopt the use of data-driven models to answer reservoir management questions. Reservoir simulation and simple analytical methods remain in use, but a new type of model has recently gained popularity to answer questions related to well spacing and targeting or optimal completion design and practices.

These new data analytics efforts usually gather data on many wells and try to determine the key factors driving economic performance (LaFollette and Holcomb 2011, 2012; Portis et al. 2013). Data-driven models are being used at different stages of the life cycle of unconventional reservoirs. First-pass models are often created to support transactional decisions during acquisition or divestiture projects to assess and rank the potential value of various basins, specific land positions, or entire companies. Some models are built early in the development of a play to support the appraisal efforts by quantifying uncertainties and methodically reducing them through data acquisition or field pilot efforts organized through ED. Today, most operators involved in unconventional reservoirs maintain evergreen models that help optimize their development efforts. These models help answer questions related to drilling and completion designs and practices or well targeting and spacing.

This section presents a general approach followed by applications for developing data-driven models for unconventional reservoirs (Courtier et al. 2016; Wicker et al. 2016, 2017; Burton et al. 2019). Published references will be provided when possible, but several companies using data analytics to model unconventional reservoirs are reluctant to publish details concerning the structure or the impact of their work because of the proprietary nature of the data sets and the relative novelty of the methods, which are still evolving. The remainder of this section describes sequentially the various steps usually taken to develop a data-driven model for an unconventional asset: data collection, attribute modeling, data reduction, machine learning, and model distribution.

4.7.1. Data Collection. The first step in building a data analytics model is to gather the data relevant to the effort. This typically presents a challenge for models pertaining to reservoir management and more specifically to unconventional reservoirs because these models integrate data from a variety of disciplines and operators. A best practice is to establish from the start of the project a data pipeline, which identifies the original source of each data point and keeps track of how the data are transformed through the analytics modeling process. This allows models to be updated quickly and automatically as new data become available.

The data are usually pulled from different systems and integrated in a unique location for analysis. Some companies have developed a dedicated analytics database (Burton et al. 2019), others load their data into third party analytics platforms, but the approach can be as simple as a shared file repository where the data pulled from different systems are aggregated.

4.7.1.1. Rock Properties. The reservoir rock properties are usually extracted from a geologic model covering the area of interest. Standard geomodeling workflows are used to define the stratigraphic and structural framework and to populate the model with petrophysical properties. Seismic surveys, petrophysical logs, and core data are integrated to estimate key rock properties such as bulk volume, rock types and facies, mineral volumes, porosity, and permeability. Additional properties are often modeled that were not usually estimated in conventional reservoirs. Wicker et al. (2016) report aggregating factors such as brittleness, sweetness, and curvature that would typically not be used in a conventional modeling workflow. The natural fracture system is typically characterized, either through volume-average properties or through a discrete fracture network. Key geomechanical properties are estimated along with the regional stress field. A variety of other seismic attributes are also routinely proposed as potential indicators of well performance. The resulting geologic model usually presents itself as a series of 2D maps or as a 3D geocellular model or property volume (Courtier et al. 2016; Wicker et al. 2016, 2017).

4.7.1.2. Fluid Properties. Most unconventional reservoirs are source rocks that still contain the organic material that formed the hydrocarbon fluids in place. A proper characterization of the fluid system therefore should encompass both an understanding of the kerogen and its maturity and a determination of the phase behavior and flow properties of the gaseous and liquid hydrocarbons. These two aspects of the fluid system are typically addressed sequentially.

The question of the kerogen and its maturity is usually tackled first, as part of the geologic modeling process. This is typically done by analysis of cuttings or core samples and petrophysical logs (Ahmed and Meehan 2016). The output of that work is usually a series of attributes providing insights into the hydrocarbon source system, such as the total organic carbon and vitrinite reflectance. Additional attributes are also routinely included that might include geochemical tracers. This effort provides an estimated distribution of the hydrocarbon fluids present in the reservoir.

Further characterization of the produced fluids is then performed to establish the phase behavior and flow characteristics of the fluids. Water properties such as salinity, dissolved chemical compositions, or pH are usually determined from produced-water samples and mapped. Produced-oil and -gas samples are frequently analyzed in terms of specific gravity, GORs, and condensate/gas ratios, and PVT experiments are conducted on a subset of the samples. Different approaches are taken for hydrocarbon fluid characterization, ranging from the calibration of an EOS to the use published correlations. A common practice is to first map simple fluid parameters (e.g., gravity measurements, geochemistry, or produced fluid ratios) to model the fluid compositions across the reservoir. The PVT samples are then used to establish correlations between these basic mapped properties and a more complete set, including the estimate evolution of these properties with respect to pressure and temperature. Black-oil-type properties such as formation volume factors, solution ratios, or viscosities can thus be estimated at any location in the field (Burton et al. 2019).

4.7.1.3. Drilling and Completion Data. Information about the well design and condition is often included in the analysis. Drilling and completion design as well as operational execution parameters are aggregated on a well-by-well or stage-by-stage basis.

On the drilling side, the well trajectory is usually processed to provide insights into the completed well length, orientation of the well in azimuth and inclination

(Wicker et al. 2017), and tortuosity. Measurements taken during drilling operations, such as measurement while drilling or mud logs, are sometimes included.

The completion data set is usually carefully compiled. When building a model from public data sources, the only information that is consistently available might be limited to the total volume of fluid and proppant injected during the fracturing process. When available, a much richer data set is included, that will include the type of fluid and proppant being used, the number of stages, the types and number of perforations. The type of completion (e.g., plug and perforate, sliding sleeves, or other) is also often included. Pump curve data are typically processed to yield key information describing the fracture design and the reservoir response (Temizel et al. 2015). A set of attributes is built that typically includes the leakoff, breakdown, propagation, instantaneous shut-in, and closure pressures, along with the average and maximum treating rates and the timing of the injection.

4.7.1.4. Lift Design and Production Operations. Various artificial lift methods are used in unconventional plays. The methods depend on a variety of factors, including depth, pressure gradient, temperature, expected water cut, and GOR. Including the artificial lift system design and its operational parameters can be very critical to help the model explain differences between wells with different lift conditions. The data can be included in several ways. When the lift systems across all wells in a study are standardized, feeding the operational parameters of the lift methods might be sufficient. When a more diverse set of lift systems exist, it might be beneficial to estimate the flowing bottomhole pressures of each well and feed that information to the model. This allows the analytics model to better quantify the influence of changing lift conditions with a relatively restricted set of input parameters.

Precomputing the flowing bottomhole pressures for wells has the added benefit of accounting for different flowback strategies. In some basins, different flowback strategies have an influence on the recovery of the wells. This can easily be accounted for by feeding the model with the flowing-bottomhole-pressure measurements or estimates, which will account for the choke settings, temperature, pressures, fluid properties, and other phenomena.

4.7.1.5. Multiwell Effects. During the development of unconventional wells in North America, it is sometimes necessary to place an initial well in a section to hold the lease by production. Operators who have acquired a large land position often initiate a program where a single well is completed in each drilling section. This initial development is then followed by an infill development where additional wells are drilled in each section. The performance of the children wells is usually lower than that of the parent wells because of depletion effects. It is therefore important to include parameters that account for the available drainage volume of a well along with the amount of depletion that occurred around the well before its completion.

4.7.1.6. Production Performance. Most data-driven models built for unconventional assets aim at forecasting future well production (Deen et al. 2013). It is therefore critical to gather the historical performance of the well in terms of oil, water, and gas production volumes. Different approaches are taken to prepare the production data set. Sometimes, the data set is compiled monthly, often in terms of cumulative production. At other times, the available production data on a given well are fit to a simplified model described by a few parameters. These parameters then form a summarized description of the well behavior, which can be fed to the analytics algorithm.

A common challenge for these models is the fact that each well has a different life span. The data set available to make a prediction therefore decreases as the forecasting time frame increases. This can lead to a statistical phenomenon called "survivor bias" where only the wells with the longest histories are used for the long-term forecast. The usual way to resolve this challenge is to append a simple forecast to the existing well production, to extend the data set. This appended forecast is usually provided through a decline curve model or using a type curve calibrated to the early well production (Fetkovich 1973; Mohaghegh 2017; Mohaghegh et al. 2017; Burton et al. 2019).

4.7.1.7. Economic Parameters. Most development optimization problems come down to finding the optimal return on investment for a given well location and design. Historical well costs and commodity prices are thus gathered along with assumptions on future economic parameters. These data are rarely fed directly to the analytics model; they are most often used in a separate economics model that uses the production forecasts to estimate the financial performance of the wells (Burton et al. 2019).

4.7.2. Machine Learning. A variety of machine learning models have been applied to forecast the production of unconventional wells. This includes common algorithms, training, and scenario analysis and optimization.

4.7.2.1. Common Algorithms. The simplest, and among the most frequently used, machine learning model is a multivariate linear regression (Courtier et al. 2016; Wicker et al. 2016, 2017; Burton et al. 2019). Although the model itself is linear, the preliminary data transformation allows for a model that is effectively nonlinear. A preprocessing algorithm known as alternating conditional expectation is sometimes used to identify potential data transformation that leads to a superior fit. More-advanced regression algorithms such as multivariate adaptive regressive splines have been used successfully over the past few years.

More-advanced models are also routinely used. Neural networks have always been a favored choice because of their versatility. For the most part, the oil industry has so far been using simple feed-forward back-propagation algorithms for these applications. These models perform well, but iterations on the network architecture are sometimes necessary, which makes these models less attractive than simpler algorithms.

Random forest is another frequently used algorithm for these problems, but it tends to be poor at extrapolation. Some companies have modified random forest algorithms to improve the extrapolation capabilities by accounting for expected parameter trends, but in general these models should be used carefully outside of the existing data range. Other algorithms such as support vector machines, Bayesian belief networks, boosted trees, or convolutional networks have also been used, although less frequently.

4.7.2.2. Scenario Analysis and Optimization. Data analytics models for unconventional reservoirs are often used by asset team members for scenario analysis.

The team tests various completion designs, well spacing or targeting strategies, or flowback or artificial lift approaches and assesses their potential merit in terms of recovery and economic performance (Burton et al. 2019). Sometimes the optimal scenarios are identified more rigorously by coupling the analytics models to an optimization engine. Wicker et al. (2017) have reported using Monte Carlo analysis on their multivariate analysis model to better understand the individual influence of key design parameters.

Although the use of data analytics to model unconventional reservoirs is relatively widespread among operators, the exact impact of these efforts is rarely publicized. The benefits are usually categorized either in work efficiency gains or in direct improvements to the development strategy. For example, an operating company (Devon) was reported (Jacobs 2016) to have dramatically improved their analysis efficiency through data-driven models. It was reported that an analysis that used to take 1 week to be performed on 50 wells was accelerated and expanded to take less than 10 minutes for the lower 48 states. Laredo has reported that their model represented a fundamental driver for value creation within field development planning and quantified the potential effect of their work by showing that higher or lower production attributes could lead to a production range of 75 to 130% around their type curve.

Summary

- *Today, because of the scale and pace of unconventional developments, data analytics is used by most operators to help forecast and optimize new unconventional-field wells and drilling units.*
- *Algorithms that have reportedly been used successfully include multivariate regression and random forest.*

5. Future Trends

Oil and gas operations are becoming increasingly more complex; with infrastructures dramatically expanding and more-sophisticated systems used both offshore and onshore. Environmental, safety, and regulatory pressure also continues to rise. Additionally, new workforce challenges are emerging as part of the big crew change.

In response, companies are preparing for sustained low prices and competitive landscape with the success of unconventionals by embracing innovative technologies led by data analytics and adopting external trends. Several reservoir engineering advancements are being led through data-driven insights and subsurface analytics.

At the basic level, certain repetitive jobs involving data migration, data processing, and event detection are being automated, freeing up reservoir engineers to perform more analysis and interpretations. In this case, automation augments reservoir engineers doing more with less.

5.1. Data. According to a recent report by Forrester, less than 0.5% of all data collected is ever analyzed. Just a 10% increase in data accessibility will result in more than USD 65 million additional net income for a typical Fortune 1000 company. Oil and gas companies have started to invest heavily in digitization, data acquisition, and data ownership to transform the business and realize the value proposition as seen in other industries. This heavy investment in big data and analytics is expected to continue, resulting in companies aggregating huge volumes of data. This further requires focus on improving data accessibility and usability (Gantz and Reinsel 2012; McKenzie et al. 2016).

Democratic and easy access to data will promote transparency and lead to quick insights, and a single version of truth will break silos across organizations and promote accountability. However, exploding data volumes will need open standard big data architecture that can seamlessly work across multiple vendor applications and operator workflows. Companies are looking into building data platforms (including subsurface data lakes) upon which multiple systems interact. This would also require standards and open interfaces for application systems to work successfully. Best practices in data profiling, data transformation, and data governance are extremely critical for sustained value creation.

There is also a shift in paradigm from moving data to calculations to moving calculations to data. This is enabled by massively parallel processing capabilities with newer hardware (CPUs and GPUs) and stream computing. The shift toward platform-as-a-service and software-as-a-service models enabling engineering and geoscience work processes will require large amounts of data to seamlessly transfer between multiple systems.

However, subsurface data are typically fraught with uncertainties, while interpretations are time-consuming and are characterized by low volumes. This is a current limitation for machine learning and AI methods as applied to reservoir engineering applications. In some cases, data labeling and cost of subsurface data acquisition form a bottleneck for generating large training data sets for supervised learning. Therefore, there is a significant potential for hybrid models, reduced-physics methods, and unsupervised learning methods in reservoir engineering applications.

5.2. Field Automation. For intelligent fields, there is a growing need for tools that enable engineers to "see" farther into the reservoir in order to adjust the flow control strategy as required. With sensors becoming less costly, more widespread, and more connected, oil and gas fields are becoming more instrumented. Downhole sensors in wells and flowlines are becoming more common, to measure pressure, temperature, multiphase flow rates, and fluid density. Novel measurement methods such as fiber optics (distributed acoustic sensing and distributed temperature sensing) have made possible measurements that offer higher spatial resolution and data frequency than traditional sensors (Smolem and van der Spek 2003).

From a reservoir management perspective, the greatest value can be achieved when implementing intelligent flow control systems coupled with surveillance tools. A robust surveillance strategy is fundamentally enabled by resilient data acquisition systems and active flow control systems that facilitate optimization of recovery efficiency.

Standardized end devices and segregation of the automation network will provide a sound platform for the flourishment of internet of things (IoT) in the oil field. Recent advancements in IoT devices and edge analytics will allow engineers to collect more data and run their algorithms in real time near the devices to optimize operations. Further, the intersection of cloud and mobility is creating new opportunities for engineers to collaborate and access relevant data from virtually anywhere and at any time. An increasing number of reservoir surveillance and simulation systems are exploring cloud-based mobile applications, which adds new pressure on IT to consider how to integrate these with modern data architecture in a secure manner.

In cases where direct measurement of certain variables is not feasible, data-driven virtual measurements are filling in the gap. Digital twins are virtual or simulated models of a physical asset. The asset could be a process, a product, or even a service. A digital twin is built by capturing data from diverse sources including the sensors installed on the different parts of the physical object and building a virtual model to track the physical asset state using fit-for-purpose data-driven and physical equations. While the concept of the digital twin is not new, the development of IoT and data analytics is making it a viable alternative for continuous monitoring (Poddar 2018).

5.3. Applications. A desirable objective in reservoir management and field optimization is to use fit-for-purpose, automated tools to transform toward data-driven advisory systems recommending decisions under uncertainty, potentially leading toward autonomous closed-loop control. Improved understanding of complex reservoir processes and enabling technology makes this shift possible from heuristics toward data-driven decisions. These trends can be seen currently in unconventional and EOR applications.

HPC advances have enabled creating and running larger and complex models that run faster with rapid advancements in GPUs and parallel computing. With increased availability of training data, the new hardware technology also enables faster learning. Coupled with intelligent algorithms and reduced-physics modeling approaches, this can speed up reservoir modeling and forecasting. Faster turnaround times for interpretations and large-scale reservoir evaluations can help optimize field development decisions. Quicker identification of issues through automated surveillance will help improve production efficiency and recovery factors.

Current applications of machine learning methods can be mostly classified as generic algorithms applied on oil-and-gas data. However, work is ongoing to start designing custom algorithms designed specifically for oil-and-gas problems. These new algorithms should be far superior to the standard out-of-the-box methods for applications such as seismic processing and interpretation and production forecasting.

Various approaches have been used to integrate physics into machine learning algorithms, but a new generation of algorithms is now doing it very naturally. The algorithms can transition smoothly between interpolation in a data-rich environment to physics-based extrapolation when needed. This approach also has the potential for replacing standard reservoir simulation solvers because the inherent architecture of the machine learning algorithm is more easily parallelizable than the standard numerical methods for partial-differential equations.

Machine learning strategies based on reinforcement learning might be adapted for reservoir engineering by including both synthetic and real data. For example, for optimal development plans or well locations, similar models might be developed in which real examples and simulation results are used together in the training. Our industry, therefore, should be more flexible in terms of data sharing.

Rooted in statistics, data analytics can also address uncertainty to make robust decisions. In particular, unconventional-field development has the potential to be strongly driven by data analytics, with large well density, fast decision cycles, and shared digital information. It is expected that more-cognitive AI applications will be used to search for hidden trends across structured and unstructured field data combining, for example, time-series analysis, text mining, and audio and video analytics.

However, while fit-for-purpose physics-based models can often address a variety of questions (within reason, as long as the phenomena are captured by the modeled physics), data-driven models have narrow focus and solve specific problems that they have been trained for. It is not surprising to envision several focused data analytics models that solve highly focused problems in place of an integrated asset model. Also, there is a strong need for more research on assessing model validity of data-driven models automatically over longer periods of time. Subsurface processes are highly nonlinear and time-varying, and the validity range of data-driven models often has to be refreshed through continuous training. However, it is necessary to distinguish between data uncertainty and model uncertainty for a sustainable process.

5.4. People. One of the severe impediments to reaping the benefits of data analytics in reservoir engineering applications is the shortage of practitioners with hybrid skill sets—combining domain expertise, computing skills, and data analytics. Companies and universities are starting to realize the need for this new breed of engineers by establishing new innovation centers.

Companies will also have to design new career development paths for these reservoir engineers so that these roles are seen as an integral part of doing business and as fundamentally transforming the business. Up- skilling current employees and democratizing data are essential to alleviate the impact of any skills gap. Current employees can be trained internally or externally. There is a wealth of training programs available that can provide them with data science skills to supplement the business experience. However, often they lack industry specific examples, business context, application idiosyncrasies, and detailed know-how. Larger oil and gas companies can also set up analytics centers of excellence to ensure that more-traditional engineers receive some training in the use of data. Industry forums (such as SPE) can also play an important role in improving the awareness and raising the skills level through training and certification programs.

Different levels of expertise might be needed to successfully establish the benefits of data-driven methods. Basic awareness programs will provide a foundation for managers and decision makers. Individuals who will be responsible for analysis of data analytic products will need to be skilled in basic techniques, and understand their strengths and weaknesses to review work done by others. Practitioners tasked with developing data products will benefit from advanced training with hands-on experience.

Academic institutions are starting to cater to these new industry skills through certificate programs, specialized courses, and cross-disciplinary curriculum combining data science and scientific computing. Massive open online courses with practically unlimited participation have focused on data science skills that have been popular in recent years. A few digital petroleum engineering programs have been established in select schools in response to these new industry demands.

Gartner estimates that over 40% of data science tasks will be automated by 2020, resulting in increased productivity and broader usage of data and analytics by citizen data scientists. This will further increase the reach of new technologies across the enterprise as well as help overcome the skills gap. Automation will simplify tasks that are repetitive and manually intensive. Increasingly, oil and gas companies have renewed their interests in implementing digital operations platforms that cater to these needs.

So, what can automation reasonably achieve in the next few years?

- Data—Companies that value data as an asset are pursuing integration of data sources to provide clean and high-quality data for data-driven methods. Automated pipelines for data engineering and data processing will limit tedious manual efforts and allow management by exception. However, this does not eliminate the need of practitioners for exploratory data analysis, assessing data needs, and making sense of the data.
- Models—Another key cognitive task related to model selection and search for architectures is also being addressed. Leading technology companies and open source initiatives in data analytics have been investing heavily in developing automated machine learning and neural architecture search frameworks. They aim to provide a suite of machine learning tools that will allow easy training of high performance deep neural networks (Hinton et al. 2012), without requiring specialized knowledge of deep learning or AI. Other major efforts are in the area of interpretability through explainable AI that can help deliver meaningful insights for decision-making purposes. Some of the biggest advances are yet to come through discovering more-robust training methods.

The authors are optimistic about what the future holds for the role of data analytics in reservoir engineering and the oil and gas industry in general. Just 10 years ago, only a handful of colleges in the US offered big data or analytics degree programs. Today, more than 100 schools have data-related undergraduate and graduate degrees, as well as certificates for working professionals or graduate students wanting to augment other degrees. The popular demand for these courses portrays a healthy pipeline of job candidates that will be realized over time. The confluence of these new hybrid engineers with modern skill sets, increasing data availability, and demonstration of more successful applications of data analytics will make these methods part of routine business processes instead of optional methods.

6. References

Aanonsen, S. I., Naevdal, G., Oliver, D. S. et al. 2009. The Ensemble Kalman Filter in Reservoir Engineering—A Review. *SPE J.* **14** (3): 393–412.

Abbeel, P., Quigley, M., and Ng, A. 2006. Using Inaccurate Models in Reinforcement Learning. Presented at the In Proceedings of the 23rd International Conference on Machine Learning, Pittsburgh, Pennsylvania, USA, 25–29 June.

Abdel-Aal, R. E. 2002. Abductive Networks: A New Modeling Tool for the Oil and Gas Industry. Presented at the SPE Asia Pacific Oil and Gas Conference and Exhibition, Melbourne, Australia, 8–10 October. SPE-77882-MS. https://doi.org/10.2118/77882-MS. https://doi.org/10.2118/77882-MS.

Adeeyo, Y. A. 2016. Artificial Neural Network Modeling of Bubblepoint Pressure and Formation Volume Factor at Bubblepoint Pressure of Nigerian Crude Oil. Presented at the SPE Nigeria Annual International Conference and Exhibition, Lagos, Nigeria, 2–4 August. SPE-184378-MS. https://doi.org/10.2118/184378-MS. https://doi.org/10.2118/184378-MS.

Adeyemi, O. T. S., Shryock, S. G., Sankaran, S. et al. 2008. Implementing "I Field" Initiatives in a Deepwater Green Field, Offshore Nigeria. Presented at the SPE Annual Technical Conference and Exhibition, Denver, Colorado, USA, 21–24 September. SPE-115367-MS. https://doi.org/10.2118/115367-MS.

Afra, S. and Tarrahi, M. 2015. Assisted EOR Screening Approach for CO_2 Flooding with Bayesian Classification and Integrated Feature Selection Techniques. Presented at the Carbon Management Technology Conference, Sugar Land, Texas, 17–19 November. SPE-440237-MS. https://doi.org/10.7122/440237-MS.

Agarwal, R. G., Al-Hussainy, R., and Ramey, H. J. 1970. An Investigation of Wellbore Storage and Skin Effect in Unsteady Liquid Flow: I. Analytical Treatment. *SPE J.* **10** (3): 279–290.

Ahmed, U. and Meehan, N. 2016. *Unconventional Oil and Gas Resources: Exploitation and Development,* first edition. Boca Raton, FL: CRC Press.

Al-Anazi, A. F. and Gates, I. D. 2010. Support-Vector Regression for Permeability Prediction in a Heterogeneous Reservoir: A Comparative Study. *SPE J.* **13** (3): 485–495.

Alakbari, F. S., Elkatatny, S., and Baarimah, S. O. 2016. Prediction of Bubble Point Pressure Using Artificial Intelligence AI Techniques. Presented at the SPE Middle East Artificial Lift Conference and Exhibition, Manama, Kingdom of Bahrain, 30 November–1 December. SPE-184208-MS. https://doi.org/10.2118/184208-MS.

Alarfaj, M. K., Abdulraheem, A., and Busaleh, Y. R. 2012. Estimating Dewpoint Pressure Using Artificial Intelligence. Presented at the SPE Saudi Arabia Section Young Professionals Technical Symposium, Dhahran, Saudi Arabia, 19–21 March. SPE-160919-MS. https://doi.org/10.2118/160919-MS.

Albertoni, A. and Lake, L. W. 2003. Inferring Inter-well Connectivity Only From Well-Rate Fluctuations in Waterfloods. *SPE J.* **6** (1): 6–16. SPE-126339-PA. https://doi.org/10.2118/83381-PA.

Alboudwarej, H. and Sheffield, J. M. 2016. Development and Application of Probabilistic Fluid Property (PVT) Models for Stochastic Black Oil Reservoir Simulation. Presented at the SPE Western Regional Meeting, Anchorage, Alaska, USA, 23–26 May. SPE-180383-MS. https://doi.org/10.2118/180383-MS.

Ali, T. A. and Sheng, J. J. 2015. Production Decline Models: A Comparison Study. Presented at the SPE Eastern Regional Meeting, 13–15 October. SPE-177300-MS. https://doi.org/10.2118/177300-MS.

Alimadadi, F., Fakhri, A., Farooghi, D. et al. 2011. Using a Committee Machine With Artificial Neural Networks To Predict PVT Properties of Iran Crude Oil. *SPE J.* **14** (1): 129–137.

Alimonti, C. and Falcone, G. 2004. Integration of Multiphase Flow Metering, Neural Networks, and Fuzzy Logic in Field Performance Monitoring. *SPE J.* **19** (1): 25–32.

Alvarado, V., Ranson, A., Hernandez, K. et al. 2002. Selection of EOR/IOR Opportunities Based on Machine Learning. Presented at the European Petroleum Conference, Aberdeen, United Kingdom, 29–31 October. SPE-78332-MS. https://doi.org/10.2118/78332-MS.

Amirian, E., Leung, J. Y. W., Zanon, S. D. J. et al. 2014. An Integrated Application of Cluster Analysis and Artificial Neural Networks for SAGD Recovery Performance Prediction in Heterogeneous Reservoirs. Presented at the SPE Heavy Oil Conference-Canada, Calgary, Alberta, Canada, 10–12 June. SPE-170113-MS. https://doi.org/10.2118/170113-MS.

Amudo, C., Xie, J., Pivarnik, A. et al. 2014. Application of Design of Experiment Workflow to the Economics Evaluation of an Unconventional Resource Play. Presented at the SPE Hydrocarbon Economics and Evaluation Symposium, Houston, Texas, 19–20 May. SPE-169834-MS. https://doi.org/10.2118/169834-MS.

Anifowose, F., Ewenla, A., and Eludiora, S. I. 2011. Prediction of Oil and Gas Reservoir Properties Using Support Vector Machines. Presented at the International Petroleum Technology Conference, Bangkok, Thailand, 15–17 November. IPTC-14514-MS. https://doi.org/10.2523/IPTC-14514-MS.

Archenaa, J. and Mary Anita, E. A. 2015. A Survey of Big Data Analytics in Healthcare and Government. *Procedia Comput Sci.* **50** (2015): 408–413.

Arief, I. H., Forest, T., and Meisingset, K. K. 2017. Estimating Fluid Properties Using Surrogate Models and Fluid Database. Presented at the SPE Europec, London, UK, 3–6 June. SPE-185937-MS. https://doi.org/10.2118/185937-MS.

Armacanqui T., J. S., Eyzaguirre G., L. F., Prudencio B., G. et al. 2017. Improvements in EOR Screening, Laboratory Flood Tests and Model Description to Effectively Fast Track EOR Projects. Presented at the Abu Dhabi International Petroleum Exhibition and Conference, Abu Dhabi, UAE, 13–16 November. SPE-188926-MS. https://doi.org/10.2118/188926-MS.

Arps, J. J. 1945. Analysis of Decline Curves. In *Transactions of the Society of Petroleum Engineers,* Vol. 160, Number 1, 228–247. Richardson, Texas: SPE. SPE-945228-G. https://doi.org/10.2118/945228-G.

Artus, V., Houze, O., and Chen, C.-C. 2019. Flow Regime-Based Decline Curve for Unconventional Reservoirs: Generalization to Anomalous Diffusion and Power Law Behavior. Presented at the Unconventional Resources Technology Conference, Denver, Colorado, USA, 22–24 July. URTEC-2019-293-MS. https://doi.org/10.15530/urtec-2019-293.

Aulia, A. and Ibrahim, M. I. 2018. DCA-Based Application for Integrated Asset Management. Presented at the Offshore Technology Conference Asia, Kuala Lumpur, Malaysia, 20–23 March. OTC-28308-MS. https://doi.org/10.4043/28308-MS.

Ballin, P. R., Shirzadi, S., and Ziegel, E. 2012. Waterflood Management Based on Well Allocation Factors for Improved Sweep Efficiency: Model Based or Data Based? Presented at the SPE Western Regional Meeting, Bakersfield, California, USA, 21–23 March. SPE-153912-MS. https://doi.org/10.2118/153912-MS.

Bandyopadhyay, P. 2011. Improved Estimation of Bubble Point Pressure of Crude Oils: Modeling by Regression Analysis. Presented at the SPE Annual Technical Conference and Exhibition, Denver, Colorado, USA, 30 October–2 November. SPE-152371-STU. https://doi.org/10.2118/152371-STU.

Batycky, R. P. Blunt, M. J., and Thiele, M. R. 1997. A 3D Field-Scale Streamline-Based Reservoir Simulator. *SPE Res Eng* **12** (4): 246–254.

Bertocco, R. and Padmanabhan, V. 2014. Big Data Analytics in Oil and Gas. *Bain,* 26 March 2014, https://www.bain.com/insights/big-data-analytics-in-oil-and-gas (accessed 18 April 2019).

Bestagini, P., Lipari, V., and Tubaro, S. 2017. A Machine Learning Approach to Facies Classification Using Well Logs. Presented at the 2017 SEG International Exposition and Annual Meeting, Houston, Texas, 24–29 September. SEG-2017-17729805. https://doi.org/10.1190/segam2017-17729805.1.

Bhark, E. and Dehghani, K. 2014. Assisted History Matching Benchmarking: Design of Experiments-Based Techniques. Presented at the SPE Annual Technical

Conference and Exhibition, Amsterdam, The Netherlands, 27–29 October. SPE-170690-MS. https://doi.org/10.2118/170690-MS.

Bond, B. N. and Daniel, L. 2008. Guaranteed Stable Projection-Based Model Reduction for In-definite and Unstable Linear Systems. Presented at the IEEE/ACM International Conference on Computer-Aided Design, San Jose, California, USA, 10–13 November. ICCAD-2008-4681657. https://doi.org/10.1109/ICCAD.2008.4681657.

Box, G.E.P. and Jenkins, G. 2008. *Time Series Analysis, Forecasting and Control*, fourth edition. Hoboken, New Jersey: Wiley.

Brown, J. B., Salehi, A., Benhallam, W. et al. 2017. Using Data-Driven Technologies to Accelerate the Field Development Planning Process for Mature Field Rejuvenation. Presented at the SPE Western Regional Meeting, Bakersfield, California, USA, 23 April. SPE-185751-MS. https://doi.org/10.2118/185751-MS.

Bruce, W. A. 1943. An Electrical Device for Analyzing Oil-Reservoir Behavior. *J Pet Technol* 51 (1): 112–124.

Brulé, M. R. 2015. The Data Reservoir: How Big Data Technologies Advance Data Management and Analytics in E&P. Presented at the SPE Digital Energy Conference and Exhibition, The Woodlands, Texas, USA, 3–5 March. SPE-173445-MS. https://doi.org/10.2118/173445-MS.

Budennyy, S., Pachezhertsev, A., Bukharev, A. et al. 2017. Image Processing and Machine Learning Approaches for Petrographic Thin Section Analysis. Presented at the SPE Russian Petroleum Technology Conference, Moscow, Russia, 16–18 October. SPE-187885-MS. https://doi.org/10.2118/187885-MS.

Burton, M., Matringe, S., Atchison, T. et al. 2019. A Data-Driven Modeling Methodology to Support Unconventional Reservoir Development Decisions: Application to the STACK Play in Oklahoma. Presented at the SPE/AAPG/SEG Unconventional Resources Technology Conference, Denver, Colorado, USA, 22–24 July. URTEC-2019-573-MS. https://doi.org/10.15530/urtec-2019-573.

Cao, F., Luo, H., and Lake, L. W. 2014. Development of a Fully Coupled Two-Phase Flow Based Capacitance Resistance Model (CRM). Presented at the SPE Improved Oil Recovery Symposium, Tulsa, Oklahoma, USA, 12–16 April. SPE-169485-MS. https://doi.org/10.2118/169485-MS.

Cao, F., Luo, H., and Lake, L.W. 2015. Oil-Rate Forecast by Inferring Fractional-Flow Models from Field Data with Koval Method Combined with the Capacitance/Resistance Model. *SPE Reservoir Evaluation & Engineering* (SPE-173315 originally presented in 2015 *SPE Reservoir Simulation Symposium*).

Cardoso, A., Durlofsky, L. J, and Sarma, P. 2009. Development and Application of Reduced-Order Modeling Procedures for Subsurface Flow Simulation. *Int J Numer Meth Eng* 77 (9): 1322–1350.

Cardoso, M. A. and Durlofsky, L. J. 2010. Linearized Reduced-Order Models for Subsurface Flow Simulation. *J. Comput Phys* 229 (3): 681–700.

Carter, R. 1985. Type Curves for Finite Radial and Linear Flow System. *SPE J.* 25 (5): 719–728.

Castellini, A., Gross, H., Zhou, Y. et al.2010. An Iterative Scheme to Construct Robust Proxy Models. Presented at the ECMOR XII – 12th European Conference on the Mathematics of Oil Recovery, Oxford, UK, 6–9 September. EAGE-2214-4609.20144999. https://doi.org/10.3997/2214-4609.20144999.

Cerchiello, P. and Giudici, P. Big Data Analysis for Financial Risk Management. *J. Big Data* 3: 18. https://doi.org/10.1186/s40537-016-0053-4.

Chatfield, C. 1996. Model Uncertainty and Forecast Accuracy, *J. Forecasting* **15** (7): 495–508.

Chaturantabat, S. and Sorensen, D. C. 2009. Discrete Empirical Interpolation for Nonlinear Model Reduction. Presented at the Joint 48th Conference on Decision and Control, Shanghai, China, 15–18 December. IEEE-CDC.2009.5400045. https://doi.org/10.1109/CDC.2009.5400045.

Chen, H., Klie, H., and Wang, Q. 2013. A Black-Box Interpolation Method To Accelerate Reservoir Simulation Solutions. Presented at the SPE Reservoir Simulation Symposium, The Woodlands, Texas, USA, 18–20 February. SPE-163614-MS. https://doi.org/10.2118/163614-MS.

Chen, B., He, J., Wen, X.-H. et al. 2017. Uncertainty Quantification and Value of Information Assessment Using Proxies and Markov Chain Monte Carlo Method for a Pilot Project. *J Petrol Sci Eng* **157**: 328–339.

Ciezobka, J., Courtier, J., and Wicker, J. 2018. Hydraulic Fracturing Test Site (HTFS) – Project Overview and Summary of Results. Presented at the SPE/AAPG/SEG Unconventional Resources Technology Conference, Houston, Texas, USA, 23–25 July. URTEC-2937168-MS. https://doi.org/10.15530/URTEC-2018-2937168.

Courtier, J., Wicker, J., and Jeffers, T. 2016. Optimizing the Development of a Stacked Continuous Resource Play in the Midland Basin. Presented at the SPE/AAPG/SEG Unconventional Resources Technology Conference, San Antonio, Texas, USA, 1–3 August. URTEC-2461811-MS. https://doi.org/10.15530/ URTEC-2016-2461811.

Cunningham, C. F., Wozniak, G., Cooley L. et al. 2012. Using Multiple Linear Regression To Model EURs of Horizontal Marcellus Wells. Presented at the SPE Eastern Regional Meeting, Lexington, Kentucky, USA, 3–5 October. SPE-161343-MS. https://doi.org/10.2118/161343-MS.

Danesh, A. 1998. *PVT and Phase Behaviour of Petroleum Reservoir Fluids,* first edition. Amsterdam: Elsevier.

Da Silva, M. and Marfurt, K. 2012. Framework for EUR Correlation to Seismic Attributes in the Barnett Shale, TX. Presented at the 2012 SEG Annual Meeting, Las Vegas, Nevada, 4–9 November. SEG-2012-1601. https://doi.org/10.1190/ segam2012-1601.1.

Das, O., Aslam, M., Bahuguna, R. et al. 2009. Water Injection Monitoring Techniques For Minagish Oolite Reservoir In West Kuwait. Presented at the International Petroleum Technology Conference, Doha, Qatar, 7 December. IPTC-13361-MS. https://doi.org/10.2523/IPTC-13361-MS.

Datta-Gupta, A. and King, M. J. 2007. *Streamline Simulation: Theory and Practice.* Richardson, Texas: Society of Petroleum Engineers.

David, R. M. 2016. Approach Towards Establishing Unified Petroleum Data Analytics Environment to Enable Data Driven Operations Decisions. Presented at the SPE Annual Technical Conference and Exhibition, Dubai, UAE, 26–28 September. SPE-181388-MS. https://doi.org/10.2118/181388-MS.

Deen, T., Shchelokov, V., Wydrinski, R. et al. 2013. Horizontal Well Performance Prediction Early in the Life of the Wolfcamp Oil Resources Play in the Midland Basin. Presented at the SPE/AAPG/SEG Unconventional Resources Technology Conference, Denver, Colorado, USA, 12–14 August. URTEC-1582281-MS. https://doi.org/10.1190/urtec2013-081.

Deng, L. and Dong, Y. 2014. Deep Learning: Methods and Applications, Foundations and Trends in Signal Processing. The Netherlands: Now Publishing.

Denney, D. 2002. Multicriteria Decision-Making in Strategic Reservoir Planning. *J Pet Technol* **54** (9): 83–84

Deutsch, C. V. and Journel, A. G. 1992. *GSLIB: Geostatistical Software Library and User's Guide,* second edition. New York, New York: Oxford University Press.

Dindoruk, B. and Christman, P. G. 2001. PVT Properties and Viscosity Correlations for Gulf of Mexico Oils. Presented at the SPE Annual Technical Conference and Exhibition, New Orleans, Louisiana, 30 September–3 October. SPE-71633-MS. https://doi.org/10.2118/71633-MS.

Duong, A. 2010. An Unconventional Rate Decline Approach for Tight and Fracture-Dominated Gas Wells. Presented at the 2010 Canadian Unconventional Resources and International Petroleum Conference, Calgary, Alberta, Canada, 19–21 October. SPE-137748-MS. https://doi.org/10.2118/137748-MS.

Dursun, S. and Temizel, C. 2013. Efficient Use of Methods, Attributes, and Case-Based Reasoning Algorithms in Reservoir Analogue Techniques in Field Development. Presented at the SPE Digital Energy Conference, The Woodlands, Texas, USA, 5–7 March. SPE-163700-MS. https://doi.org/10.2118/163700-MS.

Dzurman, P. J., Leung, J., Zanon, S. D. J. et al. 2013. Data-Driven Modeling Approach for Recovery Performance Prediction in SAGD Operations. Presented at the SPE Heavy Oil Conference – Canada, Calgary, Alberta, Canada 11–13 June. SPE-165557-MS. https://doi.org/10.2118/165557-MS.

Elsharkawy, A. M. 1998. Modeling the Properties of Crude Oil and Gas Systems Using RBF Network. Presented at the SPE Asia Pacific Oil and Gas Conference and Exhibition, Perth, Australia, 12–14 October. SPE-49961-MS. https://doi.org/10.2118/49961-MS.

El-Sebakhy, E. A., Sheltami, T., Al-Bokhitan, S. Y. et al. 2007. Support Vector Machines Framework for Predicting the PVT Properties of Crude Oil Systems. Presented at the SPE Middle East Oil and Gas Show and Conference, Manama, Bahrain, 11–14 March. SPE-105698-MS. https://doi.org/10.2118/105698-MS.

Emerick, A. A. and Reynolds, A. C. 2013. Ensemble Smoother with Multiple Data Assimilation *Comput. Geosci.* Vol. 55, 3–15.

Evensen, G. 2004. Sampling Strategies and Square Root Analysis Schemes for the EnKF. *Ocean Dynam* **54** (6): 539–560.

Fernandez, D., Vidal, J. M., and Festini, D. E. 2012. Decision-Making Breakthrough Technology. Presented at the SPE Latin America and Caribbean Petroleum Engineering Conference, Mexico City, Mexico, 16–18 April. SPE-152524-MS. https://doi.org/10.2118/152524-MS.

Fetkovich, M. J. 1973. Decline Curve Analysis Using Type Curves, AIME 48 Annual Fall Meeting, Las Vegas, Nevada. SPE-4629-MS. https://doi.org/10.2118/4629-MS.

Fetkovich, M. J. 1980. Decline Curve Analysis Using Type Curves. *J Pet Technol* **32** (6): 1065–1077. SPE-4629-PA. https://doi.org/10.2118/4629-PA.

Friedman, J. H. 1991. Multivariate Adaptive Regression Splines. *Ann Stat* Vol. 19, Number 1, 1–67.

Friedmann, F., Chawathe, A., and Larue, D. K. 2003. Assessing Uncertainty in Channelized Reservoir Using Experimental Design. *SPE Res Eval & Eng* **6** (4): 264–274. SPE-85117-PA. https://doi.org/10.2118/85117-PA.

Gautschi, W. 1994. Algorithm 726: ORTHPOL—A Package of Routines for Generating Orthogonal Polynomials and Gauss-Type Quadrature Rules. *ACM Trans Math Softw.* Vol. 20, Number 1, 21–62.

Gentil, P. H. 2005. *The Use of Multilinear Regression Models in Patterned Water-floods: Physical Meaning of the Regression Coefficients.* MS thesis, The University of Texas at Austin, Austin, Texas (August 2005).

Ghasemi, M., Yang, Y., Gildin, E. et al. 2015. Fast Multiscale Reservoir Simulations using POD-DEIM Model Reduction. Presented at the SPE Reservoir Simulation Symposium, Houston, Texas, USA, 23–25 February. SPE-173271-MS. https://doi.org/10.2118/173271-MS.

Ghoraishy, S. M., Liang, J.-T., Green, D. W. et al. 2008. Application of Bayesian Networks for Predicting the Performance of Gel-Treated Wells in the Arbuckle Formation, Kansas. Presented at the SPE Symposium on Improved Oil Recovery, Tulsa, Oklahoma, USA, 20–23 April. SPE-113401-MS. https://doi.org/10.2118/113401-MS.

Gladkov, A., Sakhibgareev, R., Salimov, D. et al. 2017. Application of CRM for Production and Remaining Oil Reserves Reservoir Allocation in Mature West Siberian Waterflood Field. Presented at the SPE Russian Petroleum Technology Conference, Moscow, Russia, 16–18 October. SPE-187841-MS. https://doi.org/10.2118/187841-MS.

Glaso, O. 1980. Generalized Pressure-Volume-Temperature Correlations. *J Pet Technol* **32** (5): 785–795. SPE-8016-PA. https://doi.org/10.2118/8016-PA.

Goldstein, P. 2018. Utilities Will Invest Heavily in Data Analytics in the Years Ahead. *biztechmagazine,* 5 January 2018, https://biztechmagazine.com/article/2018/01/utilities-will-invest-heavily-data-analytics-years-ahead (accessed 18 April 2018).

Golub, G. H. and Van Loan, C. F. 1996. *Matrix Computations,* third edition. Baltimore, Maryland: The Johns Hopkins University Press.

Gomez, Y., Khazaeni, Y., Mohaghegh, S. D. et al. 2009. Top Down Intelligent Reservoir Modeling. Presented at the SPE Annual Technical Conference and Exhibition, New Orleans, Louisiana, USA, 4–7 October. SPE-124204-MS. https://doi.org/10.2118/124204-MS.

Goodfellow, I., Bengio, Y., and Courville, A. 2016. *Deep Learning,* first edition. Cambridge, Massachusetts: MIT Press.

Grebe, M., Rüßmann, M., Leyh, M. et al. 2018. Digital Maturity Is Paying Off. *bcg,* 7 June 2018, https:// www.bcg.com/en-us/publications/2018/digital-maturity-is-paying-off.aspx (accessed 18 April 2018).

Grønnevet, M., Barnett, M., and Semenov, A. 2015. Converting Science Into Decision Making Support. Presented at the OTC Arctic Technology Conference, Copenhagen, Denmark, 23–25 March. OTC-25493-MS. https://doi.org/10.4043/25493-MS.

Guerillot, D. R. 1988. EOR Screening With an Expert System. Presented at the Petroleum Computer Conference, San Jose, California, USA, 27–29 June. SPE-17791-MS. https://doi.org/10.2118/17791-MS.

Guille, C. and Zech, S. 2016. How Utilities Are Deploying Data Analytics Now. *bain,* 30 August 2016, https://www.bain.com/insights/how-utilities-are-deploying-data-analytics-now/ (accessed 18 April 2018).

Guo, Z., Chen, C., Gao, G. et al. 2017. EUR Assessment of Unconventional Assets Using Machine Learning and Distributed Computing Techniques. Presented at the SPE/AAPG/SEG Unconventional Resources Technology Conference, Austin, Texas, USA, 24–26 July. URTEC-2659996-MS. https://doi.org/10.15530/URTEC-2017-2659996.

Guo, Z., Reynolds, A. C. and Zhao, H. 2018. A Physics-Based Data-Driven Model for History-Matching, Prediction and Characterization of Waterflooding Performance. *SPE J.* **23** (2): 367–395.

Guo, Z. and Reynolds, A. C. 2019. INSIM-FT-3D: A Three-Dimensional Data-Driven Model for History Matching and Waterflooding Optimization. Presented at the SPE Reservoir Simulation Conference, Galveston, Texas, USA, 10–11 April. SPE-193841-MS. https://doi.org/10.2118/193841-MS.

Gupta, S., Saputelli, L. A., Verde, A. et al. 2016. Application of an Advanced Data Analytics Methodology to Predict Hydrocarbon Recovery Factor Variance Between Early Phases of Appraisal and Post-Sanction in Gulf of Mexico Deep Offshore Assets. Presented at the Offshore Technology Conference, Houston, Texas, USA, 2–5 May. OTC-27127-MS. https://doi.org/10.4043/27127-MS.

Güyagüler, B., Horne, R. N., Rogers, L. et al. 2000. Optimization of Well Placement in a Gulf of Mexico Waterflooding Project. Presented at the SPE Annual Technical Conference and Exhibition, Dallas, Texas, 1–4 October. SPE-63221-MS. https://doi.org/10.2118/63221-MS.

Hanafy, H. H., Macary, S. M., ElNady, Y. M. et al. 1997. Empirical PVT Correlations Applied to Egyptian Crude Oils Exemplify Significance of Using Regional Correlations. Presented at the International Symposium on Oilfield Chemistry, Houston, Texas, 18–21 February. SPE-37295-MS. https://doi.org/10.2118/37295-MS.

Hand, D. J. 2007. Mining Personal Banking Data to Detect Fraud. In *Selected Contributions in Data Analysis and Classification,* ed. P. Brito P., G. Cucumel, G. Bertrand P. et al. 377–386. Berlin, Heidelberg: Springer.

Harding, T. B. 1996. Life Cycle Value/Cost Decision Making. Presented at the International Petroleum Conference and Exhibition of Mexico, Villahermosa, Mexico, 5–7 March. SPE-35315-MS. https://doi.org/10.2118/35315-MS.

Hayashi, A. 2013. Thriving in a Big Data World. *MIT Sloan Management Review* **55** (2): 35-39.

He, J. 2010. *Enhanced Linearized Reduced-Order Models for Subsurface Flow Simulation.* MS thesis, Stanford University, Stanford, California (June 2010).

He, J. and Durlofsky, L. J. 2010. Use of Linearized Reduced-Order Modeling and Pattern Search Methods for Optimization of Oil Production. Presented at the 2nd International Conference on Engineering Optimization, Lisbon, Portugal, 6–9 September.

He, J., Sarma, P. and Durlofsky, L. 2011a. Use of Reduced-Order Model for Improved Data Assimilation Within an ENKF Context. Presented at the SPE Reservoir Simulation Symposium, The Woodlands, Texas, USA, 21–23 February. SPE-141967-MS. https://doi.org/10.2118/141967-MS.

He, J., Satrom, J., and Durlofsky, L. J. 2011b. Enhanced Linearized Reduced-Order Models for Subsurface Flow Simulation. *J. Comput. Phys.* **230** (23): 8313–8341.

He, J., Sarma, P., and Durlofsky, L. J. 2013. Reduced-Order Flow Modeling and Geological Parameterization for Ensemble-Based Data Assimilation. *Comput Geosci* **55**: 54–69.

He, J. 2013. Reduce-Order Modeling for Oil-Water and Compositional Systems, with Application to Data Assimilation and Production Optimization. PhD dissertation, Stanford University, Stanford, California (October 2013).

He, J., Xie, J, Wen, W. and Wen, X.-H. 2015a. Improved Proxy For History Matching Using Proxy-for-Data Approach and Reduced Order Modeling. Presented at the SPE Western Regional Meeting, Garden Grove, California, USA, 27–30 April. SPE-174055-MS. https://doi.org/10.2118/174055-MS.

He, J., Xie, J., Sarma, P. et al. 2015b. Model-Based A Priori Evaluation of Surveillance Programs Effectiveness Using Proxies. Presented at the SPE Reservoir Simulation Symposium, Houston, Texas, USA, 23–25 February. SPE-173229-MS. https://doi.org/10.2118/173229-MS.

He, J., Xie, J., Wen, X.H. et al. 2016a. An Alternative Proxy for History Matching Using Proxy-For-Data Approach and Reduced Order Modeling. *J Petrol Sci Eng* **146**: 392–399.

He, J., Xie, J., Sarma, P. et al. 2016b. Proxy-Based Workflow for a Priori Evaluation of Data-Acquisition Programs. *SPE J.* **21** (4): 1400–1412.

He, J., Sarma, P., Bhark, E. et al. 2017a. Quantifying Value of Information Using Ensemble Variance Analysis. Presented at the SPE Reservoir Simulation Conference, Montgomery, Texas, USA, 20–22 February. SPE-182609-MS. https://doi.org/10.2118/182609-MS.

He, J., Tanaka, S., Wen, X.H. et al. 2017b. Rapid S-Curve Update Using Ensemble Variance Analysis with Model Validation. Presented at the SPE Western Regional Meeting, Bakersfield, California, 23–27 April. SPE-185630-MS. https://doi.org/10.2118/185630-MS.

He, J., Sarma, P., Bhark, E. et al. 2018. Quantifying Expected Uncertainty Reduction and Value of Information Using Ensemble-Variance Analysis. *SPE J.* **23** (2): 428–448. SPE-182609-PA. https://doi.org/ 10.2118/182609-PA.

Hinton, G. et al., "Deep Neural Networks for Acoustic Modeling in Speech Recognition: The Shared Views of Four Research Groups," *IEEE Signal Processing Magazine,* vol. 29, no. 6, pp. 82–97, Nov. 2012.

Hoeink, T. and Zambrano, C. 2017. Shale Discrimination with Machine Learning Methods. Presented at the 51st U.S. Rock Mechanics/Geomechanics Symposium, San Francisco, California, USA, 25–28 June. ARMA-2017-0769.

Holanda, R. W. de, Gildin, E., and Jensen, J. L. 2015. Improved Waterflood Analysis Using the Capacitance- Resistance Model Within a Control Systems Framework. Presented at the SPE Latin American and Caribbean Petroleum Engineering Conference, Quito, Ecuador, 18–20 November. SPE-177106-MS. https://doi.org/10.2118/177106-MS.

Holanda, R. W. de, Gildin, E., & Valkó, P. P. 2017. Combining Physics, Statistics and Heuristics in the Decline Curve Analysis of Large Datasets in Unconventional Reservoirs. Society of Petroleum Engineers. SPE-185589-MS. https://doi.org/10.2118/185589-MS.

Holanda, R.W., Gildin, E., Jensen, J.L. et al. 2018. A State-of-the-Art Literature Review on Capacitance Resistance Models for Reservoir Characterization and Performance Forecasting. *Energies* **11** (12): 3368.

Holdaway, K. 2014. Harness Oil and Gas Big Data with Analytics: Optimize Exploration and Production with Data-Driven Models, first edition. Hoboken, New Jersey: Wiley & Sons.

Ilk, D., Rushing, J. A., Perego, A. D. and Blasingame, T. A. 2008. Exponential vs. Hyperbolic Decline in Tight Gas Sands: Understanding the Origin and Implications for Reserve Estimates Using Arps' Decline Curves. Presented at the SPE Annual Technical Conference and Exhibition at Denver, USA. 21–24 September. SPE-116731-MS. https://doi.org/10.2118/116731-MS.

Jacobs, T. 2016. Devon Energy Rises to the Top as a Data-Driven Producer. *J Pet Technol* **68** (10): 28–29. SPE-1016-0028 JPT. https://doi.org/10.2118/1016-0028-JPT.

Jafarizadeh, B. and Bratvold, R. B. 2009. Strategic Decision Making in the Digital Oil Field. Presented at the SPE Digital Energy Conference and Exhibition, Houston, Texas, USA, 7–8 April. SPE-123213-MS. https://doi.org/10.2118/123213-MS.

Javadi, F. and Mohaghegh, S. D. 2015. Understanding the Impact of Rock Properties and Completion Parameters on Estimated Ultimate Recovery in Shale. Presented at the SPE Eastern Regional Meeting, Morgantown, West Virginia, USA, 13–15 October. SPE-177318-MS. https://doi.org/10.2118/177318-MS.

Journel, A. G. and Huijbregts, C. J. 1978. *Mining Geostatistics,* 7th edition. London: Academic Press.

Kalantari Dahaghi, A. and Mohaghegh, S. D. 2009. Top-Down Intelligent Reservoir Modeling of New Albany Shale. Presented at the SPE Eastern Regional Meeting, Charleston, West Virginia, USA. 23–25 September. SPE-125859-MS. https://doi.org/10.2118/125859-MS.

Kansao, R., Yrigoyen, A., Haris, Z. et al. 2017. Waterflood Performance Diagnosis and Optimization Using Data-Driven Predictive Analytical Techniques from Capacitance Resistance Models (CRM). Presented at the 79th EAGE Annual Conference and Exhibition, Paris, France 12 June. SPE-185813-MS. https://doi.org/10.2118/185813-MS.

Kaushik, A., Kumar, V., Mishra, A. et al. 2017. Data Driven Analysis for Rapid and Credible Decision Making: Heavy Oil Case Study. Presented at the Abu Dhabi International Petroleum Exhibition & Conference, Abu Dhabi, UAE, 13–16 November. SPE-188635-MS. https://doi.org/10.2118/188635-MS.

Klie, H. 2013. Unlocking Fast Reservoir Predictions via Nonintrusive Reduced-Order Models. Presented at the SPE Reservoir Simulation Symposium, The Woodlands, Texas, USA, 18–20 February. SPE-163584-MS. https://doi.org/10.2118/163584-MS.

Klie, H. 2015. Physics-Based and Data-Driven Surrogates for Production Forecasting. Presented at the   SPE Reservoir Simulation Symposium, Houston, Texas, USA, 23–25 February. SPE-173206-MS. https://doi.org/10.2118/173206-MS.

LaFollette, R., Holcomb, W. 2011. Practical Data Mining: Lessons Learned from the Barnett Shale of North Texas. Presented at the SPE Hydraulic Fracturing Technology Conference, 24–26 January, The Woodlands, Texas, USA. SPE-140524-MS. https://doi.org/10.2118/140524-MS.

LaFollette, R., Holcomb, W. 2012. Practical Data Mining: Analysis of Barnett Shale Production Results with Emphasis on Well Completion and Fracture Stimulation. Presented at the SPE Hydraulic Fracturing Technology Conference, 6–8 February, The Woodlands, Texas, USA. SPE-152531-MS. https://doi.org/10.2118/152531-MS.

Landa, J.L. and Güyagüler, B. 2003. A Methodology for History Matching and the Assessment of Uncertainties Associated with Flow Prediction. Presented at the SPE Annual Technical Conference & Exhibition, Denver, Colorado, 5–8 October. SPE-84465-MS. https://doi.org/10.2118/84465-MS.

Lasater, J. A. 1958. Bubble Point Pressure Correlation. *J Pet Technol* **10** (5): 65–67. SPE-957-G. https://doi.org/ 10.2118/957-G.

Lee, J., Rollins, J.B. and Spivey J.P. 2003, *Pressure Transient Testing*. SPE Textbook Series, Vol. 9. ISBN: 978-1-55563-099-7.

Lerlertpakdee, P., Jafarpour, B. and Gildin, E. 2014. Efficient Production Optimization With Flow-Network Models. *SPE J*, 19 (6):1083–1095.

Le Maitre, O.P., Kino, O., Najm, H. et al. 2001. A Stochastic Projection Method for Fluid Flow. *J. Comput. Phys* 173 (2): 481–511.

Li, B. and Friedmann, F. 2005. Novel Multiple Resolution Design of Experiments and Response Surface Methodology for Uncertainty Analysis of Reservoir Simulation Forecasts. Presented at the SPE Reservoir Simulation Symposium, The Woodlands, Texas, 31 January–2 February. SPE-92853-MS. https://doi.org/10.2118/92853-MS.

Li, H., Sarma, P. and Zhang, D. 2011. A Comparative Study of the Probabilistic Collocation and Experimental Design Methods for Petroleum Reservoir Uncertainty Quantification. *SPE J*. 16 (2): 429–439. SPE-140738-PA. https://doi.org/10.2118/140738-PA.

Li, H. and Zhang, D. 2007. Probabilistic Collocation Method for Flow in Porous Media: Comparisons With Other Stochastic Methods. *Water Resour Res*. 43 (9): 1–13. https://doi.org/10.1029/2006WR005673.

Li, H. and Zhang, D. 2009. Efficient and Accurate Quantification of Uncertainty for Multiphase Flow With the Probabilistic Collocation Method. *SPE J*. 14 (4): 665–679. SPE-114802-PA. https://doi.org/ 10.2118/114802-PA.

Lin, J., de Weck, O.L., and MacGowan, D. 2012. Modeling Epistemic Subsurface Reservoir Uncertainty Using a Reverse Wiener Jump–Diffusion Process. *J Petrol Sci Eng* 84–85: 8–19. https://doi.org/10.1016/ j.petrol.2012.01.015.

Lino, A., Vyas, A., Huang, J., Datta-Gupta, A., Fujita, Y., Bansal, N. and Sankaran, S. 2017a. Efficient Modeling and History Matching of Shale Oil Reservoirs Using the Fast Marching Method: Field Application and Validation. Presented at the SPE Western Regional Meeting, Bakersfield, California, 23–27 April. SPE-185719-MS. https://doi.org/10.2118/185719-MS.

Lino, A., Vyas, A., Huang, J., Datta-Gupta, A., Fujita, Y. and Sankaran, S. 2017b. Rapid Compositional Simulation and History Matching of Shale Oil Reservoirs Using the Fast Marching Method. Presented at the SPE/AAPG/SEG Unconventional Resources Technology Conference, Austin, Texas, USA, 24–26 July. URTEC-2693139-MS. https://doi.org/10.15530/URTEC-2017-2693139.

Lu, X., Sun, S., and Dodds, R. 2016. Toward 70% Recovery Factor: Knowledge of Reservoir Characteristics and IOR/EOR Methods from Global Analogs. Presented at the SPE Improved Oil Recovery Conference, Tulsa, Oklahoma, USA, 11–13 April. SPE-179586-MS. https://doi.org/10.2118/179586-MS.

Lutidze, G. and Younis, R. 2018. Proxy Reservoir Simulation Model for IOR Operations. Presented at the 8th Saint Petersburg International Conference & Exhibition, Saint Petersburg, Russia, 9–12 April. https://doi.org/10.3997/2214-4609. 201800284.

Ma, Z., Liu, Y., Leung, J., Zanon, S. 2015. Practical Data Mining and Artificial Neural Network Modeling for SAGD Production Analysis, Society of Petroleum Engineers, SPE Canada Heavy Oil Technical Conference, Calgary, Alberta, Canada, 9–11 June. SPE-174460-MS. https://doi.org/10.2118/174460-MS.

Maqui, A. F., Zhai, X., Negreira, S., Matringe, S.F. 2017. A Comprehensive Workflow for Near Real Time Waterflood Management and Production Optimization Using Reduced-Physics and Data-Driven Technologies. Society of Petroleum Engineers, SPE Latin America and Caribbean Petroleum Engineering

Conference, 17–19 May, Buenos Aires, Argentina. SPE-185614-MS. https://doi.org/10.2118/185614-MS.

Market Research Engine. 2018. Digital Oilfield Market By Process Analysis (Reservoir Optimization, Drilling Optimization, Production Optimization); By Solutions Analysis (Hardware Solutions, Software & Service Solutions, Data Storage Solutions) and by Regional Analysis—Global Forecast by 2017–2023. Report ID: EMDOM318, Market Research Engine, Deerfield Beach, Florida (March 2018).

McCain, W. D., Soto, R. B., Valkó, P. P. et al. 1998. Correlation of Bubblepoint Pressures for Reservoir Oils—A Comparative Study. Presented at the SPE Eastern Regional Meeting, Pittsburgh, Pennsylvania, 9–11 November. SPE-51086-MS. https://doi.org/10.2118/51086-MS.

McCain, W.D. 1990. The Properties of Petroleum Fluids. Tulsa, Oklahoma: PennWell Publishing Co. McKenzie, W., Schave, R., Farnan, M., Deny, L., Morrison, P., and Hollingsworth, J. 2016. A New Communications Protocol for Real-Time Decision Making. Society of Petroleum Engineers. SPE-181088-MS. https://doi.org/10.2118/181088-MS.

Mehta, A. 2016. Tapping the Value from Big Data Analytics. *J Pet Technol* **68** (12): 40–41. SPE-1216-0040JPT. https://doi.org/10.2118/1216-0040-JPT.

Mishra, K. K. 2017. Selecting Forecasting Methods in Data Science. *datasciencecentral,* 13 February 2017, https://www.datasciencecentral.com/profiles/blogs/selecting-forecasting-methods-in-data-science (accessed on 10 May 2018).

Mishra, S and Datta-Gupta, A. 2017. *Applied Statistical Modeling and Data Analytics: A Practical Guide for the Petroleum Geosciences,* first edition. Amsterdam, The Netherlands: Elsevier.

Mohaghegh, S., Arefi, R., Ameri, S. et al. 1995. Design and Development of an Artificial Neural Network for Estimation of Formation Permeability. *SPE Computer Applications* **7** (6): 151–154. SPE-28237-PA. https://doi.org/10.2118/28237-PA.

Mohaghegh S., Modavi C., Hafez H. et al. 2006. Development of Surrogate Reservoir Models (SRM) For Fast Track Analysis of Complex Reservoirs. Presented at the Intelligent Energy Conference and Exhibition, Amsterdam, The Netherlands, 11–13 April. SPE-99667-MS. https://doi.org/10.2118/99667-MS.

Mohaghegh, S. D., Al-Mehairi, Y., Gaskari, R. et al. 2014. Data-Driven Reservoir Management of a Giant Mature Oilfield in the Middle East. Presented at the SPE Annual Technical Conference and Exhibition, Amsterdam, The Netherlands, 27–29 October. SPE-170660-MS. https://doi.org/10.2118/170660-MS.

Mohaghegh, S. D. 2016. Top-Down Modeling: A Shift in Building Full-Field Models for Mature Fields. *J Pet Technol.* July 2016.

Mohaghegh, S. D., Gaskari, R., and Maysami, M. 2017. Shale Analytics: Making Production and Operational Decisions Based on Facts: A Case Study in Marcellus Shale. Presented at the SPE Hydraulic Fracturing Technology Conference and Exhibition, The Woodlands, Texas, USA, 24–26 January. SPE-184822-MS. https://doi.org/10.2118/184822-MS.

Mohaghegh, S. 2017. *Shale Analytics: Data-Driven Analytics in Unconventional Resources,* first edition. New York, New York: Springer.

Mohammadmoradi, P. and Kantzas, A. 2018. Wettability and Capillary Imbibition in Shales; Analytical and Data-Driven Analysis. Presented at the SPE Canada Unconventional Resources Conference, Calgary, Alberta, Canada, 13–14 March. SPE-189806-MS. https://doi.org/10.2118/189806-MS.

Montgomery, D. C. 2012. *Design and Analysis of Experiments,* eighth edition. Hoboken, New Jersey: Wiley Publishing Company.

Molinari, D., Sankaran, S., Symmons, D. et al. 2019a. Implementing an Integrated Production Surveillance and Optimization System in an Unconventional Field. Presented at the Unconventional Resources Technology Conference, Denver, USA, 22–24 July. URTEC-2019-41. https://doi.org/10.15530/URTEC-2019-41.

Molinari, D., Sankaran, S., Symmons, D. et al. 2019b. A Hybrid Data and Physics Modeling Approach Towards Unconventional Well Performance Analysis. Presented at the SPE Annual Technology Conference & Exhibition, Calgary, Canada, 30 September–2 October. SPE-196122-MS. https://doi.org/10.2118/196122-MS.

Møller, J. B., Meisingset, K. K., Arief, I. H. 2018. An Improved Correlation Approach to predict Viscosity of Crude Oil Systems on the NCS. Presented at the SPE Norway One Day Seminar, Bergen, Norway, 18 April. SPE-191296-MS. https://doi.org/10.2118/191296-MS.

Moradi, B., Malekzadeh, E.,Amani, M. et al. 2010. Bubble Point Pressure Empirical Correlation. Presented at the Trinidad and Tobago Energy Resources Conference, Port of Spain, Trinidad, 27–30 June. SPE-132756-MS. https://doi.org/10.2118/132756-MS.

Moussa, T., Elkatatny, S., Abdulraheem, A. et al. 2017. A Hybrid Artificial Intelligence Method to Predict Gas Solubility and Bubble Point Pressure. Presented at the SPE Kingdom of Saudi Arabia Annual Technical Symposium and Exhibition, Dammam, Saudi Arabia, 24–27 April. SPE-188102-MS. https://doi.org/10.2118/188102-MS.

Muhammad U.S., Muhammad K.H., Ramzan T. et al. "A Survey of Big Data Analytics in Healthcare" *International Journal of Advanced Computer Science and Applications(ijacsa),* 8 (6), 2017. http://dx.doi.org/10.14569/IJACSA.2017.080646.

Negara, A., Jin, G., and Agrawal, G. 2016. Enhancing Rock Property Prediction from Conventional Well Logs Using Machine Learning Technique—Case Studies of Conventional and Unconventional Reservoirs. Presented at the Abu Dhabi International Petroleum Exhibition & Conference, Abu Dhabi, UAE, 7–10 November. SPE-183106-MS. https://doi.org/10.2118/183106-MS.

Nemeth, L. K., and Kennedy, H. T. 1967. A Correlation of Dewpoint Pressure With Fluid Composition and Temperature. *SPE J.* 7 (2): 99–104. SPE-1477-PA. https://doi.org/10.2118/1477-PA.

Ng, A., 2018. Machine Learning Yearning—Technical Strategy for AI Engineers in the Era of Deep Learning. *mlyearning,* https://www.mlyearning.org (accessed 18 April 2018).

Nikravesh, M., Kovscek, A. R., Murer, A. S. et al. 1996. Neural Networks for Field-Wise Waterflood Management in Low Permeability, Fractured Oil Reservoirs. Presented at the SPE Western Regional Meeting, Anchorage, Alaska, 22–24 May. SPE-35721-MS. https://doi.org/10.2118/35721-MS.

North American CRO Council. 2017. Risk Implications of Data and Analytics to the Insurance Industry, http://www.crocouncil.org/images/CROC_Risk_Implications_of_Data_and_Analytics_to_the_Insurance_ Industry_20170106.pdf.

Okpere, A., and Njoku, C. 2014. Application of Neural Networks in Developing an Empirical Oil Recovery Factor Equation for Water Drive Niger Delta Reservoirs. 5 August. Society of Petroleum Engineers. SPE-172489-MS. https://doi.org/10.2118/172489-MS.

Okoduwa, I. G. and Ikiensikimama, S. S. 2010. Bubble point Pressure Correlations for Niger Delta Crude Oils. Presented at the Nigeria Annual International Conference and Exhibition, Tinapa—Calabar, Nigeria, 31 July–7 August. SPE-136968-MS. https://doi.org/10.2118/136968-MS.

Osman, E.-S. A. and Al-Marhoun, M. A. 2005. Artificial Neural Networks Models for Predicting PVT Properties of Oil Field Brines. Presented at the SPE Middle East Oil and Gas Show and Conference, Kingdom of Bahrain, 12–16 March. SPE-93765-MS. https://doi.org/10.2118/93765-MS.

Palacio, C. and Blasingame, T. 1993. Decline-Curve Analysis Using Type-Curves—Analysis of Gas Well Production Data. Presented at the Low Permeability Reservoirs Symposium, Denver, Colorado, 26–28 April. SPE-25909-MS. https://doi.org/10.2118/25909-MS.

Pankaj, P., Geetan, S., MacDonald, R. et al. 2018. Application of Data Science and Machine Learning for Well Completion Optimization. Presented at the Offshore Technology Conference, Houston, Texas, USA, 30 April–3 May. OTC-28632-MS. https://doi.org/10.4043/28632-MS.

Parada, C. H., and Ertekin, T. 2012. A New Screening Tool for Improved Oil Recovery Methods Using Artificial Neural Networks. Presented at the SPE Western Regional Meeting, Bakersfield, California, USA, 21–23 March. SPE-153321-MS. https://doi.org/10.2118/153321-MS.

Pedersen, K., Christensen, P., and Shaikh, J. 2014. *Phase Behavior of Petroleum Reservoir Fluids,* second edition. Boca Raton, Florida: CRC Press.

Perez-Valiente, M. L., Martin Rodriguez, H., Santos, C. N. et al. 2014. Identification of Reservoir Analogues in the Presence of Uncertainty. Presented at the SPE Intelligent Energy Conference & Exhibition, Utrecht, The Netherlands, 1–3 April. SPE-167811-MS. https://doi.org/10.2118/167811-MS.

Poddar, T. 2018. Digital Twin Bridging Intelligence Among Man, Machine and Environment. Presented at the Offshore Technology Conference Asia, Kuala Lumpur, Malaysia, 20–23 March. OTC-28480-MS. https://doi.org/10.4043/28480-MS.

Portis, D., Bello, H., Murray, M., Suliman, B., Barzola, G., Basu, N. 2013. A Comprehensive Well Look- Back; Analyzing Three Years' Worth of Drilling, Completing and Producing Eagle Ford Wells in Order to Understand Trend-Wide Performance Drivers and Variability. SPE/AAPG/SEG Unconventional Resources Technology Conference, 12–14 August, Denver, Colorado, USA. URTEC-1581570-MS. https://doi.org/10.1190/urtec2013-018.

Queipo, N. V., Goicochea P., Javier, V. et al. 2001. Surrogate Modeling-Based Optimization of SAGD Processes. Presented at the SPE International Thermal Operations and Heavy Oil Symposium, Porlamar, Margarita Island, Venezuela, 12–14 March. SPE-69704-MS. https://doi.org/10.2118/69704-MS.

Raghupathi W. and Raghupathi V. 2014. Big Data Analytics in Healthcare: Promise and Potential. *Health Inf Sci Syst* 2 (3): 1–10.

Ramirez, A. M., Valle, G. A., Romero, F. et al. 2017. Prediction of PVT Properties in Crude Oil Using Machine Learning Techniques MLT. Presented at the SPE Latin America and Caribbean Petroleum Engineering Conference, Buenos Aires, Argentina, 17–19 May. SPE-185536-MS. https://doi.org/10.2118/185536-MS.

Rassenfoss, S. 2018. Time to Enlist in the Analytics Army. *J Pet Technol* 70 (6): 33–34.

Raterman, K. T., Farrell, H. E., Mora, O. S., Janssen, A. L., Gomez, G. A., Busetti, S., McEwen, J., Davidson, M., Friehauf, K., Rutherford, J., Reid, R., Jin, G., Roy,

B., Warren, M. 2017. Sampling a Stimulated Rock Volume: An Eagle Ford Example. Presented at the SPE/AAPG/SEG Unconventional Resources Technology Conference, Austin, Texas, USA, 24–26 July. URTEC-2670034-MS. https://doi. org/10.15530/ URTEC-2017-2670034.

Reagan, M.T., Najm, H.N., Debusschere, B.J. et al. 2004. Spectral Stochastic Uncertainty Quantification in Chemical Systems. *Combust Theor Model* **8** (3): 607–632.

Reed, R. D. and Marks, R. J. 1999. *Neural Smithing: Supervised Learning in Feedforward Artificial Neural Networks,* first edition. Cambridge, Massachusetts: The MIT Press.

Ren, G., He, H., Wang, Z., Younis, R. M. and Wen, X.-H. 2019. Implementation of Physics-Based Data- Driven Models With a Commercial Simulator. Presented at the SPE Reservoir Simulation Conference, 10–11 April, Galveston, TX, USA. SPE-193855-MS. https://doi.org/10.2118/193855-MS.

Rewienski, M. and White, J. 2003. A Trajectory Piecewise-Linear Approach to Model Order Reduction and Fast Simulation of Non-Linear Circuits and Micromachined Devices. *IEEE T Comput Aid D* **22** (2):155–170. Richardson, J., Yu, W., and Weijermars, R. 2016. Benchmarking Recovery Factors of Individual Wells Using a Probabilistic Model of Original Gas in Place to Pinpoint the Good, Bad and Ugly Producers. Presented at the SPE/AAPG/SEG Unconventional Resources Technology Conference, San Antonio, Texas, USA, 1–3 August. URTEC-2457581-MS. https://doi.org/10.15530/URTEC-2016-2457581.

Rollins, B. and Herrin, M. 2015. Finding the Key Drivers of Oil Production through SAS Data Integration and Analysis. Presented at the Unconventional Resources Technology Conference, 20–22 July, San Antonio, Texas, USA. URTEC-2150079-MS. https://doi.org/10.15530/URTEC-2015-2150079.

Rousset, M., Huang, C. K., Klie, H. et al. 2014. Reduced-Order Modeling for Thermal Recovery Processes. *Comp Geo* **18** (3–4): 401–415.

Saeedi, A., Camarda, K. V., and Liang, J.-T. 2006. Using Neural Networks for Candidate Selection and Well Performance Prediction in Water-Shutoff Treatments Using Polymer Gels—A Field-Case Study. Presented at the SPE Aia Pacitic Oil & Gas Conference and Exhibition, Adelaide, Australia, 11–13 September. SPE-101028-MS. https://doi.org/10.2118/101028-MS.

Salazar-Bustamante, M., Gonzalez-Gomez, H., Matringe, S. F., and Castineira, D. (2012, January 1). Combining Decline-Curve Analysis and Capacitance/ Resistance Models To Understand and Predict the Behavior of a Mature Naturally Fractured Carbonate Reservoir Under Gas Injection. Society of Petroleum Engineers. SPE-153252-MS. https://doi.org/10.2118/153252-MS.

Sankaran, S., Lugo, J. T., Awasthi, A. et al. 2009. The Promise and Challenges of Digital Oilfield Solutions: Lessons Learned from Global Implementations and Future Directions. Presented at the SPE Digital Energy Conference and Exhibition, Houston, Texas, USA, 7–8 April. SPE-122855-MS. https://doi.org/ 10.2118/122855-MS.

Sankaran, S., Olise, M. O., Meinert, D., and Awasthi, A. 2011. Realizing Value from Implementing i-field (TM) in Agbami—A Deepwater Greenfield in an Offshore Nigeria Development. *SPE Econ & Mgmt* **3** (1): 31–44. SPE-127691-PA. https:// doi.org/10.2118/127691-PA.

Sankaran, S., Wright, D., Gamblin, H. et al. 2017. Creating Value by Implementing an Integrated Production Surveillance and Optimization System—An Operator's

Perspective. Presented at the SPE Annual Technical Conference and Exhibition, San Antonio, Texas, USA, 9–11 October. SPE-187222-MS. https://doi.org/10.2118/187222-MS.

Saputelli, L. and Nikolaou, M. 2003. Self-Learning Reservoir Management. Presented at the SPE Annual Technical Conference and Exhibition, Denver, Colorado, 5–8 October. SPE-84064-MS. https://doi.org/ 10.2118/84064-MS.

Saputelli, L. 2016. Petroleum Data Analytics. *J Pet Technol* **68** (10): 66.

Saputelli, L., Cherian, B., Gregoriadis, K. et al. 2000. Integration of Computer-Aided High-Intensity Design with Reservoir Exploitation of Remote and Offshore Locations. Presented at the International Oil and Gas Conference and Exhibition in China, Beijing, China, 7–10 November. SPE-64621-MS. https://doi.org/ 10.2118/64621-MS.

Saputelli, L., Verde, A., Haris, Z. et al. 2015. Deriving Unconventional Reservoir Predictive Models from Historic Data using Case-Based Reasoning (CBR). Presented at the Unconventional Resources Technology Conference, San Antonio, Texas, USA, 20–22 July. URTEC-2155770-MS. https://doi.org/10.15530/URTEC-2015-2155770.

Sarma, P., Fowler, G., Henning, M. et al. 2017a. Cyclic Steam Injection Modeling and Optimization for Candidate Selection, Steam Volume Optimization, and SOR Minimization, Powered by Unique, Fast, Modeling and Data Assimilation Algorithms. Presented at the SPE Western Regional Meeting, Bakersfield, California, 23–27 April. SPE-185747-MS. https://doi.org/10.2118/185747-MS.

Sarma, P., Kyriacou, S., Henning, M. et al. 2017b. Redistribution of Steam Injection in Heavy Oil Reservoir Management to Improve EOR Economics, Powered by a Unique Integration of Reservoir Physics and Machine Learning. Presented at the SPE Latin America and Caribbean Petroleum Engineering Conference, Buenos Aires, Argentina, 19 May. SPE-185507-MS, https://doi.org/10.2118/185507-MS.

Sarma, P. and Xie, J. 2011. Efficient and Robust Uncertainty Quantification in Reservoir Simulation with Polynomial Chaos Expansions and Non-intrusive Spectral Projection. Presented at the SPE Reservoir Simulation Symposium, The Woodlands, Texas, 21–23 February. SPE-141963-MS. https://doi.org/ 10.2118/141963-MS.

Sarma, P., Yang, C., Xie, J. et al. 2015. Identification of "Big Hitters" with Global Sensitivity Analysis for Improved Decision Making Under Uncertainty. Presented at the SPE Reservoir Simulation Symposium, Houston, Texas, USA, 23–25 February. SPE-173254-MS. https://doi.org/10.2118/173254-MS.

Sarwar, A. M., Hanif, M. K., Talib, R., Mobeen, A., Aslam, M. 2017. A Survey of Big Data Analytics in Healthcare. *International Journal of Advanced Computer Science and Applications* **8** (6): 355-359.

Satija A. and Caers J. 2015. Direct Forecasting of Subsurface Flow Response from Non-Linear Dynamic Data by Linear Least-Squares in Canonical Functional Principal Component Space. *Adv Water Resour* **77**: 69–81.

Sayarpour, M., Zuluaga, E., Kabir, C. S. et al. 2007. The Use of Capacitance-Resistive Models for Rapid Estimation of Waterflood Performance. Presented at the SPE Annual Technical Conference and Exhibition, Anaheim, California, USA, 11–14 November. SPE-110081-MS. https://doi.org/10.2118/110081-MS.

Sayarpour M. 2008. *Development and Application of Capacitance-Resistive Models to Water/CO₂ Floods.* PhD dissertation, University of Texas, Austin, Texas (August 2008).

Sayarpour, M., Kabir, C. S., and Lake, L. W. 2009a. Field Applications of Capacitance-Resistance Models in Waterfloods. *SPE Res Eval & Eng* **12** (6): 853–864.

Sayarpour M., Zuluaga, E., Kabir, C.S. et al. 2009b. The Use of Capacitance-Resistance Models for Rapid Estimation of Waterflood Performance and Optimization. *J Petrol Sci Eng* **69** (3–4): 227–238.

Scheidt, C., Renard, P., and Caers, J. 2015. Prediction-Focused Subsurface Modeling: Investigating the Need for Accuracy in Flow-Based Inverse Modeling. *Math. Geosci.* **47** (2): 173–191.

Shabab, M., Jin, G., Negara, A. et al. 2016. New Data-Driven Method for Predicting Formation Permeability Using Conventional Well Logs and Limited Core Data. Presented at the SPE Kingdom of Saudi Arabia Annual Technical Symposium and Exhibition, Dammam, Saudi Arabia, 25–28 April. SPE-182826-MS. https://doi.org/10.2118/182826-MS.

Shahvali, M., Mallison, B., Wei, K. et al. 2012. An Alternative to Streamlines for Flow Diagnostics on Structured and Unstructured Grids. *SPE J.* **17** (03). SPE-146446-PA. https://doi.org/10.2118/146446-PA.

Sharma, H., Mazumder, S., Gilbert, T. et al. 2013. Novel Approach to EUR estimation in Coal Seam Gas Wells. Presented at the SPE Unconventional Resources Conference and Exhibition-Asia Pacific, Brisbane, Australia, 11–13 November. SPE-167071-MS. https://doi.org/10.2118/167071-MS.

Sharma, J., Popa, A., Tubbs, D. et al. 2017. The Use of Voronoi Mapping for Production Growth in a Heavy Oil Field. Presented at the SPE Western Regional Metting, Bakersfield, California, 23–27 April. SPE-185676-MS. https://doi.org/10.2118/185676-MS.

Sidahmed, M., Ziegel, E., Shirzadi, S., Stevens, D., Marcano, M. 2014. Enhancing Wellwork Efficiency with Data Mining and Predictive Analytics. Society of Petroleum Engineers—SPE Intelligent Energy International 2014, pp. 536–548.

Sidahmed, M., Coley, C., and Shirzadi, S. 2015. Augmenting Operations Monitoring by Mining Unstructured Drilling Reports. Presented at the SPE Digital Energy Conference and Exhibition, The Woodlands, Texas, USA, 3–5 March. SPE-173429-MS. https://doi.org/10.2118/173429-MS.

Sidahmed, M. and Bailey, R. 2016. Machine Learning Approach for Irregularity Detection in Dynamic Operating Conditions. Presented at the SPE Annual Technical Conference and Exhibition, Dubai, UAE, 26–28 September. SPE-181425-MS. https://doi.org/10.2118/181435-MS.

Sidahmed, M., Roy, A., and Sayed, A. 2017. Streamline Rock Facies Classification with Deep Learning Cognitive Process. Presented at the SPE Annual Technical Conference and Exhibition, San Antonio, Texas, USA, 9–11 October. SPE-187436-MS. https://doi.org/10.2118/187436-MS.

Siena, M., Guadagnini, A., Rossa, E. D., Lamberti, A., Masserano, F., Rotondi, M. 2015. A New Bayesian Approach for Analogs Evaluation in Advanced EOR Screening. Presented at the EUROPEC 2015, Madrid, Spain, 1–4 June. SPE-174315-MS. https://doi.org/10.2118/174315-MS.

Smolem, J. and van der Spek, A. 2003. Distributed Temperature Sensing—A DTS Primer for Oil and Gas Production. energistics, May 2003, http://w3.energistics.org/schema/witsml_v1.3.1_data/doc/Shell_DTS_Primer.pdf (accessed 18 April 2018).

Solomatine, D. P. and Ostfeld, A. 2008. "Data-Driven Modelling: Some Past Experiences and New Approaches," *J. Hydroinformatics,* Vol 10, Issue 1.

Srivastava, P., Wu, X., Amirlatifi, A. and Devegowda, D. 2016. Recovery Factor Prediction for Deepwater Gulf of Mexico Oilfields by Integration of Dimensionless Numbers with Data Mining Techniques. Presented at the SPE Intelligent Energy International Conference and Exhibition, Aberdeen, UK. 6–8 September. SPE-181024-MS. https://doi.org/10.2118/181024-MS.

Srivastava, U. and Gopalkrishnan, S. 2015. Impact of Big Data Analytics on Banking Sector: Learning for Indian Banks. *Procedia Comput Sci* 50: 643–652. https://doi.org/10.1016/j.procs.2015.04.098.

Standing, M. B. 1947. A Pressure-Volume-Temperature Correlation for Mixtures of California Oils and Gases. Presented at the Drilling and Production Practice, New York, New York, 1 January. API-47-275.

Standing, M. B. 1979. A Set of Equations for Computing Equilibrium Ratios of a Crude Oil/Natural Gas System at Pressures Below 1,000 psia. *J Pet Tech* **31** (9): 1193–1195. SPE-7903-PA. https://doi.org/ 10.2118/7903-PA.

Startzman, R.A., Brummet, W.M., Ranney, J., Emanuel, A.S., and Toronyi, R.M. 1977. Computer Combines Offshore Facilities and Reservoir Forecasts. *Petroleum Engineer* May: 65–74.

Sudaryanto, B., and Yortsos, Y. C. 2001. Optimization of Displacements in Porous Media Using Rate Control.

Society of Petroleum Engineers. SPE-71509-MS. https://doi.org/10.2118/71509-MS.

Sun W., Durlofsky L., and Hui, M. 2017. Production Forecasting and Uncertainty Quantification for Naturally Fractured Reservoirs using a New Data-Space Inversion Procedure. *Computat Geosci* **21** (5–6): 1443–1458.

Surguchev, L., and Li, L. 2000. IOR Evaluation and Applicability Screening Using Artificial Neural Networks. Presented at the SPE/DOE Improved Oil Recovery Symposium, Tulsa, Oklahoma, 3–5 April. SPE-59308-MS. https://doi.org/10.2118/59308-MS.

Tan, X., Gildin, E., Trehan, S. et al. 2017. Trajectory-Based DEIM TDEIM Model Reduction Applied to Reservoir Simulation. Presented at the SPE Reservoir Simulation Conference, Montgomery, Texas, USA, 20–22 February. SPE-182600-MS. https://doi.org/10.2118/182600-MS.

Tang, D., and Spikes, K. 2017. Segmentation of Shale SEM Images Using Machine Learning. Presented at the 2017 SEG International Exposition and Annual Meeting, Houston, Texas, 24–29 September. SEG-2017-17738502.

Tarrahi, M., Afra, S., and Surovets, I. 2015. A Novel Automated and Probabilistic EOR Screening Method to Integrate Theoretical Screening Criteria and Real Field EOR Practices Using Machine Learning Algorithms. 26 October. Presented at the SPE Russian Petroleum Technology Conference, 26–28 October, Moscow, Russia. SPE-176725-MS. https://doi.org/10.2118/176725-MS.

Tatang, M. A., Pan W. W., Prinn, R. G. et al. 1997. An Efficient Method for Parametric Uncertainty Analysis of Numerical Geophysical Model. *J. Geo. Res.* **102** (D18): 21,925–21,932.

Temizel, C., Purward, S., Abdullayev, A., Urrutia, K., Tiwari, A., Erdogan, S. 2015. Efficient Use of Data Analytics in Optimization of Hydraulic Fracturing in Unconventional Reservoirs. Presented at the Abu Dhabi International Petroleum Exhibition and Conference, 9-12 November, Abu Dhabi, UAE. SPE-177549-MS. https://doi.org/10.2118/177549-MS.

Temizel, C., Salehian, M., Cinar, M. et al. 2018. A Theoretical and Practical Comparison of Capacitance- Resistance Modeling With Application to Mature Fields. Presented at the SPE Kingdom of Saudi Arabia Annual Technical Symposium and Exhibition, Dammam, Saudi Arabia, 23–26 April. SPE-192413-MS. https://doi.org/10.2118/192413-MS.

Thiele, M. R. and Batycky, R. P. 2003. Water Injection Optimization Using a Streamline-Based Workflow. Presented at the SPE Annual Technical Conference and Exhibition, Denver, Colorado, 5–8 October. SPE- 84080-MS. https://doi.org/10.2118/84080-MS.

Thiele, M. R. and Batycky, R. P. 2006. Using Streamline-Derived Injection Efficiencies for Improved Waterflood Management. *SPEREE* **9** (2). SPE-84080-PA. https://doi.org/10.2118/84080-PA.

Tian, C. and Horne, R. N. 2015. Applying Machine Learning Techniques to Interpret Flow Rate, Pressure and Temperature Data From Permanent Downhole Gauges. Presented at the SPE Western Regional Meeting, Garden Grove, California, USA, 27–30 April. SPE-174034-MS. https://doi.org/10.2118/174034-MS.

Valkó, P.P. 2009. Assigning Value to Stimulation in the Barnett Shale—A Simultaneous Analysis of 7000 plus Production Histories and Well Completion Records. Presented at the 2009 SPE hydraulic Fracturing Technology Conference, The Woodlands, TX, USA, 19–21 January. SPE-119369-MS. https://doi.org/10.2118/119369-MS.

Valkó, P. P. and McCain W. D. 2003. Reservoir Oil Bubble Point Pressures Revisited; Solution Gas–Oil Ratios and Surface Gas Specific Gravities. *J Petrol Sci Eng* **37** (3): 153–169.

Valle Tamayo, G. A., Romero Consuegra, F., Mendoza Vargas, L. F., and Osorio Gonzalez, D. A. 2017. Empirical PVT Correlations Applied for Colombian Crude Oils: A New Approach. Presented at the SPE Latin America and Caribbean Petroleum Engineering Conference, 17–19 May, Buenos Aires, Argentina. SPE-185565-MS. https://doi.org/10.2118/185565-MS.

Van Den Bosch, R. H. and Paiva, A. 2012. Benchmarking Unconventional Well Performance Predictions. Presented at the SPE/EAGE European Unconventional Resources Conference and Exhibition, Vienna, Austria, 20–22 March. SPE-152489-MS. https://doi.org/10.2118/152489-MS.

van Doren, J. F. M., Markovinovic, R., and Jansen, J. D. 2006. Reduced-Order Optimal Control of Water Flooding Using Proper Orthogonal Decomposition. *Computat Geosci* **10** (1): 137–158.

Vasquez, M., and Beggs, H. D. 1980. Correlations for Fluid Physical Property Prediction. Society of Petroleum Engineers. SPE-6719-PA. https://doi.org/10.2118/6719-PA.

Villarroel, G., Crosta, D., and Romero, C. 2017. Integration of Analytical Tools to Obtain Reliable Production Forecasts for Quick Decision-Making. Presented at the SPE Europec featured at 79th EAGE Conference and Exhibition, Paris, France, 12–15 June. SPE-185818-MS. https://doi.org/10.2118/185818-MS.

Wantawin, M, Yu, W., Dachanuwattana, S. et al. 2017. An Iterative Response-Surface Methodology by Use of High-Degree-Polynomial Proxy Models for Integrated History Matching and Probabilistic Forecasting Applied to Shale-Gas Reservoirs. *SPE J.* **22** (6): 2012–2031.

Weber D. 2009. The Use of Capacitance-Resistance Models to Optimize Injection Allocation and Well Location in Water Floods. PhD dissertation, University of Texas, Austin, TX, USA.

Wen, X.-H., Clayton V. Deutsch and Cullick, A. S. 2003. Inversion of Dynamic Production Data for Permeability: Fast Streamline-Based Computation of Sensitivity Coefficients of Fractional Flow. *J. Hydrol* **281** (4): 296–312.

Wen, X.-H and Chen, W. H. 2005. Real Time Reservoir Updating Using Ensemble Kalman Filter. Presented at the SPE Reservoir Simulation Symposium, The Woodlands, Texas, 31 January–2 February. SPE-92991-MS. https://doi.org/10.2118/92991-MS.

Wen, X.-H and Chen, W. H. 2007. Some Practical Issues on Real Time Reservoir Updating Using Ensemble Kalman Filter, *SPE J.* **12** (2): 156–166. SPE-111571-PA. https://doi.org/10.2118/111571-PA.

Whitson, C. H. and Brulé, M. R. 2000. *Phase Behavior,* Vol. 20. Richardson, Texas: Society of Petroleum Engineers.

Wicker, J., Courtier, J., and Curth, P. 2016. Multivariate Analytics of Seismic Inversion Products to Predict Horizontal Production in the Wolfcamp Formation of the Midland Basin. Presented at the SPE/ AAPG/SEG Unconventional Resources Technology Conference, San Antonio, Texas, USA, 1–3 August. URTEC-2449798-MS. https://doi.org/10.15530/URTEC-2016-2449798.

Wicker, J., Courtier, J., Gray, D., Jeffers, T., Trowbridge, S. 2017. Improving Well Designs and Completion Strategies Utilizing Multivariate Analysis. Presented at the SPE/AAPG/SEG Unconventional Resources Technology Conference, Austin, Texas, USA, 24–26 July. URTEC-2693211-MS. https://doi.org/10.15530/URTEC-2017-2693211.

Wilson, A. 2015. Creating Value With Permanent Downhole Gauges in Tight Gas Appraisal Wells. *J Pet Technol* **67** (2): 112–115. SPE-0215-0112-JPT. https://doi.org/10.2118/0215-0112-JPT.

Wilson, A. 2017. Drill and Learn: A Decision-Making Work Flow To Quantify Value of Learning. *J Pet Technol* **69** (4): 95–96. SPE-0417-0095-JPT. https://doi.org/10.2118/0417-0095-JPT.

Wuthrich, M.V. and Buser, C. 2017. Data Analytics for Non-Life Insurance Pricing. Research Paper No. 16–68, Swiss Finance Institute, Geneva, Switzerland (October 2017).

Xie, J., Gupta, N., King, M. J. et al. 2012. Depth of Investigation and Depletion Behavior in Unconventional Reservoirs Using Fast Marching Methods. Presented at the SPE Europec/EAGE Annual Conference, Copenhagen, Denmark, 4–7 June. SPE-154532-MS. https://doi.org/10.2118/154532-MS.

Xiu, D. and Karniadakis, G. E. 2002. Modeling Uncertainty in Steady State Diffusion Problems via Generalized Polynomial Chaos. *Comput Method Appl M* **191** (43): 4927–4948.

Yalgin, G., Zarepakzad, N., Artun, E. et al. 2018. Design and Development of Data-Driven Screening Tools for Enhanced Oil Recovery Processes. Society of Petroleum Engineers. Presented at SPE Western Regional Meeting, Garden Grove, California, USA, 22–26 April. SPE-190028-MS. https://doi.org/10.2118/190028-MS.

Yang, T., Basquet, R., Callejon, A. et al. 2014. Shale PVT Estimation Based on Readily Available Field Data. Presented at the SPE/AAPG/SEG Unconventional Resources Technology Conference, Denver, Colorado, 25–27 August. URTEC-1884129-MS. https://doi.org/10.15530/URTEC-2014-1884129.

Yang, T., Arief, I. H., Niemann, M., Houbiers, M. 2019a. Reservoir Fluid Data Acquisition Using Advanced Mud Logging Gas in Shale Reservoirs. Presented at the

Unconventional Resources Technology Conference, Denver, Colorado, USA, 22–24 July. URTEC-2019-383. https://doi.org/10.15530/URTEC-2019-383.

Yang, T., Arief, I. H., Niemann, M. et al. 2019b. A Machine Learning Approach to predict Gas Oil Ratio Based on Advanced Mud Gas Data. Presented at the SPE Europec, London, UK, 3–6 June. SPE-195459-MS. https://doi.org/10.2118/195459-MS.

Yang, Y., Ghasemi, M., Gildin, E. et al. 2016. Fast Multiscale Reservoir Simulations With POD-DEIM Model Reduction. *SPE J.* **21** (6): 2141–2154. SPE-173271-PA. https://doi.org/10.2118/173271-PA.

Yang, Y., Gildin, E., Efendiev, Y. et al. 2017. Online Adaptive POD-DEIM Model Reduction for Fast Simulation of Flows in Heterogeneous Media. Presented at the SPE Reservoir Simulation Conference, Montgomery, Texas, USA, 20–22 February. SPE-182682-MS. https://doi.org/10.2118/182682-MS.

Yeten, B., Castellini, A., Güyagüler, B. et al. 2005. A Comparison Study on Experimental Design and Response Surface Methodologies. Presented at the SPE Reservoir Simulation Symposium, The Woodlands, Texas, 31 January–2 February. SPE-93347-MS. https://doi.org/10.2118/93347-MS.

Yousef, A. A., Gentil, P. H., Jensen, J. L. et al. 2006. A Capacitance Model to Infer Interwell Connectivity From Production and Injection Rate Fluctuations. *SPE Res Eval & Eng* **9** (6): 630–646. SPE-95322-PA. https://doi.org/10.2118/95322-PA.

Yousef, A. A. 2006. Investigating Statistical Techniques to Infer Interwell Connectivity from Production and Injection Rate Fluctuations. PhD dissertation, University of Texas, Austin, Texas (May 2006).

Zaki, M. and Wagner, M. 2014. *Data Mining and Analysis: Fundamental Concepts and Algorithms,* first edition. Cambridge, Massachusetts: Cambridge University Press.

Zerafat, M. M., Ayatollahi, S., Mehranbod, N. et al. 2011. Bayesian Network Analysis as a Tool for Efficient EOR Screening. Presented at the SPE Enhanced Oil Recovery Conference, Kuala Lumpur, Malaysia, 19–21 July. SPE-143282-MS. https://doi.org/10.2118/143282-MS.

Zhang, Y., Bansal, N., Fujita, Y. et al. 2016. From Streamlines to Fast Marching: Rapid Simulation and Performance Assessment of Shale-Gas Reservoirs by Use of Diffusive Time of Flight as a Spatial Coordinate. *SPE J.* **21** (5): 1883–1898.

Zhang, Y., He, J., Yang, C., et al. 2018. A Physics-Based Data-Driven Model for History Matching, Prediction, and Characterization of Unconventional Reservoirs. *SPE J.* **23** (4): 1105–1125.

Zhao, H., Kang, Z., Zhang, X. et al. 2015. INSIM: A Data-Driven Model for History Matching and Prediction for Waterflooding Monitoring and Management with a Field Application. Presented at the SPE Reservoir Simulation Symposium, Houston, Texas, USA, 23–25 February. SPE-173213-MS. https://doi.org/10.2118/173213-MS.

Zhao, H., Kang, Z., Zhang, X. et al. 2016a. A Physics-Based Data-Driven Numerical Model for Reservoir History Matching and Prediction with a Field Application. *SPE J.* **21** (06): 2175–2194.

Zhao, H., Li, Y., Cui, S. et al. 2016b. History Matching and Production Optimization of Water Flooding Based on a Data-Driven Inter-Well Numerical Simulation Model. *J Nat Gas Sci Eng* **31**: 48–66.

Zhao Y. and Sarma, P. 2018. A Benchmarking Study of a Novel Data Physics Technology for Steamflood and SAGD Modeling: Comparison to Conventional Reservoir Simulation. Presented at the SPE Canada Heavy Oil Technical Conference, Calgary, Alberta, Canada, 13–14. SPE-189772-MS. https://doi.org/10.2118/189772-MS.

Appendix A—Model Development Process

Typically, the data-driven model development process involves the following steps.

1. Define the objectives and purpose for the model.
2. Analyze the system to be modeled to understand the scale, complexity, and processes involved.
3. Collect, validate, analyze, and prepare the data set available for modeling.
4. Define acceptable accuracy of the model.
5. Select model and proposed modeling approach.
6. Develop a learning model and a modeling workflow.
7. Calibrate model against held out data.
8. Validate model against blind data.
9. Deploy model for intended purpose.

Business Objectives. Every model is built with a sense of a defined purpose or objective, which should primarily drive the modeling approach. For example, when well placement optimization is the primary objective, a material balance model might not be an appropriate modeling strategy.

The business objectives could be classified as follows:

Predictive—Aimed at predicting future outcomes or what would happen. For example, predict EUR of a well from first few months of observed production (prediction problem); or predict if a water breakthrough would happen within the next 1 month (classification problem).

Explanative—Understanding why something happens by using plausible mechanisms to match outcome data in a well-defined manner. For example, explain why water breaks through prematurely in producer wells by injecting water in certain water injection wells.

Illustrative—Showing a mechanism or idea clearly to understand how it happens. For example, show that the water blocking effect is observed in flow through hydraulically fractured horizontal wells.

System Analysis. Proper understanding and delineation of the system (or the subsystems) being modeled are essential initial steps in the model scoping process. If the underlying physical mechanisms can be modeled explicitly with known model parameters in a timely manner, then this might offset the need for data-driven models. Therefore, it is important to understand the limitations or the driving factors for a data-driven model in the first place. The acceptance of a model is thus guided by its "usefulness" rather than the "truth." Having domain knowledge or partial understanding of the physics can also be helpful in designing features that can assist or speed up the machine learning model.

Data Analysis. The next phase of the process dives into a preliminary analysis of the data set to better understand the problem at hand.

- It is important to ensure reproducibility of the entire analysis that is important for data analytics projects.

Therefore, original versions of raw data must be kept separate from the cleaned-up data set (and its versions).

Exploratory Data Analysis. After the data have been gathered for the study, an initial exploratory data analysis starts, which is typically aimed at understanding the data set as much as possible before modeling it. The objective of this phase is for the modelers to familiarize themselves with each variable, understand the relationships between the variables, visualize the distributions, and identify potential gaps or outliers. During this phase, a series of descriptive statistics are calculated, and a series of displays are generated to visualize the content of the data. This important step is critical for efforts in both descriptive and diagnostic analytics. Following this initial investigation, the data set is transformed in several ways before the application of the machine learning algorithms (Mishra and Datta-Gupta 2017).

Understanding the data available for modeling starts with defining the type of data—namely, structured or unstructured. Structured data are defined by columns and data instances with well-defined data types, whether they are numerical or categorical. Unstructured data are mostly in the form of text, image, sound, or other such formats.

Data Availability. When limited data are available or the process often takes excursions outside the historical training data range, (full or reduced) physics-based models are much more reliable for better accuracy. On the other hand, when substantial relevant data are available, and the process is conducive to data-driven modeling, machine learning models can be quite useful and often outperform physical models. The choice of machine learning models is also dependent on the amount of data available for training purposes.

The size of training and validation data sets is quite important in the choice of modeling approach and meeting desired modeling accuracy levels. For example, if two different algorithms are used that have very close accuracy levels, it might not be possible to detect the difference with small training data sets. For validation data sets, a popular heuristic is to use 30% of the available data for testing purposes. In general, the availability of labeled data sets for classification applications in reservoir engineering is often limited and might be a necessity for supervised learning algorithms.

Data Types. Achieving full potential of all modeling approaches is underpinned by appropriate identification, selection, and access to data and information relevant to the study (e.g., flow transient analysis, waterflood/EOR). Data-driven, machine learning, and AI-based models for subsurface and reservoir applications operate on the following categories of data.

Time series data—Parameter values vary with time, indicating change in operating mode (through external signals) or in response to other inputs. Such data are typically input (manipulated variable such as choke position), a disturbance (e.g., noise), or a response (controlled variable such as wellhead pressure or flow rate).

In reservoir engineering, dynamic representations that have a spatial domain are an extension of time series data (i.e., a spatio-temporal system where the value depends on both location and time).

Time series data that comprise multiple parameters can be synchronous or asynchronous. Even when the measurements are overtly synchronized, they can suffer from clock errors (i.e., the clock reading might be identical, but the clocks might not be synchronized; caution is required twice a year if the time basis is local time and the clocks are adjusted to daylight savings time). Readings can be periodic or aperiodic (e.g., the process temperature might be recorded every 15 seconds while an analyzer reading might be acquired every 120 minutes and a sample might be taken once or twice a week, usually in the morning shift).

The user will need to understand the tag naming conventions, to be aware that the correct measurements are retrieved (Sidahmed et al. 2014, 2015). There might be a choice of process historian that is queried to recover the data; a mirror database might be preferred rather than the active primary historian, to avoid overloading the primary application. The user will also need to understand the units of measure of the recorded data. This ensures that the range of data (interpolation/extrapolation) is chosen appropriately for the model purpose. This is a key step to ensure model fidelity.

Stationary data (steady state)—Derived from process instrumentation and subject to electronic and process noise, drift, and other forms of calibration error, stationary data do not depend on the time that the reading is acquired because they are not changing with time.

Fixed data—These are attributes of the system rather than of the operation of the system (e.g., landing depth of a well, maximum flow coefficient of a valve).

Categorical data—The above data types are mostly numeric in nature, but certain data types can be divided into groups, commonly called categorical data. For example, a well type can be divided into vertical well, horizontal well, or slanted well; an aquifer can be classified as infinite acting, strong, weak, or no aquifer.

Data redundancy—The question of redundant information arises when identifying the tags to retrieve. Redundancy takes several forms (Sidahmed et al. 2015):

- **Temporal redundancy**—rather like the situation of stationary data, we can improve our data by a higher frequency or number of measurements taken from a single measurement tag.
- **Point measurement redundancy**—the designer intended that a single process measurement be acquired by more than one instrument, usually for the purpose of ensuring either that the measurement was acquired or that a more reliable value was recorded/acted upon through some voting system (e.g., a trip system requiring two out of three measurements to exceed a threshold, rather than having a single gauge provide a reading above the threshold value).
- **Model based redundancy**—a relationship known to be true can be exploited to provide a value or a confirmation of a value (e.g., a system known to contain a pure component can provide an estimate of a state variable such as temperature from a measurement of another state such as pressure). This can be extended to multiple measurements and multiple relationships (e.g., a whole process mass balance and a process simulation can collaborate to estimate the error in every gauge measurement).

Data may often need to be curated on the basis of a number of methods—e.g., missing value imputation, defining limits of a variable, preprocessing (scaling), feature engineering, outlier removal, variable transformation, and cross correlation.

Data cleansing—Most data-driven algorithms are sensitive to outliers and missing data. It is therefore important to prepare the data set accordingly. A first step in the analysis is often to identify entries that are clearly erroneous. This is often a straightforward statistical exercise that detects values that fall outside of a physical range or significantly outside of the distribution (called outlier detection). The problematic entries are corrected when possible or removed if necessary. Next, an effort is made to replace the missing values in the data set by best estimates (called data imputation). When possible, data points that have been removed or that were never available should be estimated to provide the model with a data set that is as complete as possible. Several approaches can be applied for such exercises, such as replacing missing values with a mean, linear interpolation, nearest-neighbor estimate, data reconciliation methods, or partial multivariate models from other available and correlated variables.

Data transformation—The type of distribution of input variables sometimes has a significant influence on the performance of machine learning algorithms. A good practice is to analyze the distribution of each variable and to investigate transformations that could normalize the data (i.e., define a transformed attribute that follows a distribution resembling a Gaussian). A classic example is to use a logarithm transform of the permeability values rather than the permeability so that the input variable is closer to a normal distribution than to a log-normal distribution. Finally, it is often recommended to normalize each variable. Although most algorithms have internal normalization algorithms, it is a good practice to prepare a data set as normalized as possible to avoid potential numerical errors.

Training, development, and test data sets—Splitting the data into training, development, and test data sets needs careful attention to the details (Ng 2018). The training data set is what the learning algorithm is run on. The development data set is used to tune parameters and select features, and is used to make other decisions regarding the learning algorithm (also called the hold-out cross-validation set). The test data set is used to evaluate the performance of the algorithm, but not to make any decisions regarding what learning algorithm or parameters to use. The training and validation (development/test) data sets should come from the same distribution.

A common heuristic is to split data into 70% for training and 30% for validation (development and test). However, when a large quantity of data is available, the test data set can be much less than 30% of the data. It is important that the development data set be large enough to detect meaningful changes in the accuracy of the algorithm.

Feature selection—Features represent the variables that are used as inputs to train a machine learning model. Feeding redundant variables to a machine learning algorithm can lead to erroneous results. Most analytics effort includes a thorough investigation of covariance between attributes. Variables containing independent information are usually retained among highly correlated variables and the redundant factors are typically discarded from the data set. This preliminary screening is most often performed and is sometimes followed by a more general data reduction step. A further reduction in the number of variables in the data set is sometimes performed through algorithms such as principal-component analysis or multidimensional scaling. This step can lead to a significant reduction in the number of factors while retaining most of the information contained in the data set. While the above methods represent unsupervised selection, it is also possible to use supervised selection methods by examining features in conjunction with a trained model where performance can be computed. This is frequently done using random forest algorithms that are well-tailored to this problem thanks to the probabilistic sampling step inherent in these methods.

Feature engineering—Depending on the type of machine learning methods used, it might be advantageous to introduce new features as a transformed variable from other input variables based on domain knowledge. For DL methods (refer to Section 3.3), features are usually simple because the algorithms generate their own transformations. However, this approach requires large amounts of data and comes at the expense of interpretability. For most other machine learning methods, feature engineering is necessary to convert data to improve performance and interpretability. These steps are important to increase the data quality and have a significant effect on model building and accuracy of the model.

Model Acceptance Criteria. It is important to establish the right model evaluation metrics before the model development process. It is recommended that this be a single-number metric that can be used to compare various models. For example, for a classification problem, using both precision (fraction of relevant instances among the retrieved instances) and recall (fraction of retrieved relevant instances over the total amount of relevant instances) would make it harder to compare algorithms. During model development, it is common to try many ideas about algorithm type, architecture, model parameters, choice of features, and other factors. However, having a single-number evaluation metric (e.g., accuracy, F1 score) allows sorting of all models according to their performance on this metric and decides quickly what is working best.

Further, the thresholds for desired accuracy and model run time (speed) should also be carefully chosen on the basis of the business objective and the available data set. Establishing optimal error rates or human level performance upfront can set reasonable expectations to assess performance of the machine learning models.

Modeling Approach. The selection of the modeling approach is determined primarily by its purpose, data availability, speed, accuracy, and interpretability requirement.

It is very difficult to know in advance what data-driven modeling approach will work best for a new problem. It is not uncommon to try out several ideas as part of this iterative process. The process involves starting off with some idea of how to build the model, implement the idea (often in code), and carry out an experiment to analyze the performance (Chatfield 1996). Therefore, the essence is in going around this loop as fast as possible. For this reason, having a good training data set and an established model acceptance criterion is very important.

Learning Model Selection. A variety of machine learning models have been applied to several reservoir engineering applications. The simplest model possible is a multivariate linear regression. Although the model itself is linear, the preliminary data transformation allows for a model that is effectively nonlinear. It is also possible to model the problem in log space, which provides a multiplicative rather than an additive model. A preprocessing algorithm known as alternating conditional expectation is sometimes used to identify potential data transformation that leads to a superior fit. More-advanced regression algorithms such as multivariate adaptive regressive splines [a nonparametric regression technique that automatically models nonlinearities and interactions between variables (Friedman 1991)] have also gained popularity over the past few years.

More-advanced models are also routinely used. Neural networks have always been a preferred choice because of their famous universal approximation capacity. For the most part, the oil industry has so far been using simple feed-forward back-propagation algorithms for these applications. These models perform well, but iterations on the network architecture are sometimes necessary, which makes these models less attractive than simpler algorithms.

Random forest is an algorithm that has gained tremendous popularity recently because of its ease of use. Random forest is essentially a model that combines the learning of an ensemble of decision trees, each decision tree providing a different classification of the data set with associated estimate for each class. The resulting algorithm is a robust nonlinear interpolation algorithm, but the model is very poor at extrapolation in its standard form. Other algorithms such as auto-regressive integrated moving average (ARIMA), finite impulse response (FIR), Box-Jenkins (BJ), support vector machines, Bayesian belief networks, boosted trees, or convolutional networks have also been used, although less frequently. Note that ARIMA, FIR, and BJ methods are applied for time series modeling.

Model Calibration. Machine learning models have the capacity to fit the data set extremely accurately. If applied without control, they can learn the data set completely and simply regurgitate the test values perfectly. Such overfitted models have little predictive accuracy. The most common way to guarantee that machine learning algorithms learn the key trends in the data sets rather than simply memorize the answer is to withhold a portion of the data for validation purposes. **Fig. A-1** illustrates the typical behavior of model complexity vs. error, where it is desirable to select the optimal model complexity to avoid underfitting or overfitting.

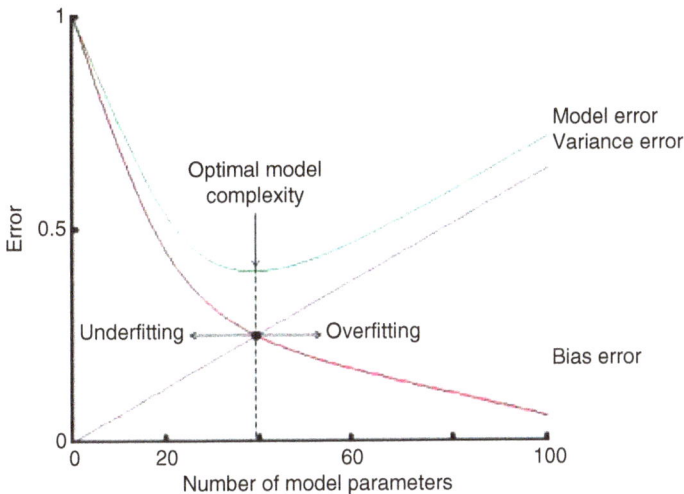

Fig. A-1—Model complexity vs. error.

Therefore, the data set is split into training, development, and test sets. As training progresses, the modeling error for the training set and the development set should decrease. When the development set error starts to increase, the model enters a phase of memorization and the training should be stopped.

Model Validation. Basic error analysis is commonly used to examine development set examples to understand the underlying causes of the errors. Bias and variance are two major sources of error in machine learning. Informally, the algorithm's error rate on the training data set is often referred to as the bias, whereas variance refers to how much worse the development (or test) set performs compared with the training set. Understanding them will help decide on a future course of action—e.g., add more data, change learning algorithm. When the error metric is mean squared error, then the total error can be expressed as a sum of bias and variance.

When high bias is found, some courses of action include

- Change model algorithm or architecture.
- Increase model size (e.g., number of layers in neural network).
- Modify input features on the basis of insights from error analysis.
- Reduce or eliminate regularization.
- Increase ranges of parameters.

When high variance is found, some courses of action include

- Add data to training set to avoid overfitting.
- Decrease model size (e.g., number of layers in neural network).
- Select features to decrease number of input features.
- Use regularization.
- Decrease ranges of parameters.

Model Deployment. Model deployment can be performed in the form of descriptive model deployment or predictive model deployment. Descriptive or diagnostic model deployment is mostly incorporating the visualization with different plotting options analyzing the data structure; basically, some visual analytics is deployed. Predictive model deployment requires a standalone model (e.g., an equation expressed as a function that operates on well-defined inputs) that is often deployed together with an application program interface and a visualization interface, and each new data set is fed into the model and the model output (predicted target such as production value) is displayed on the visualization interface.

When the models have been generated and deployed, it is often necessary to manage existing models for further analysis and historical comparison. The process requires versioning the models and saving them into a model catalog or a repository so that they can be accessed easily later.

www.ingramcontent.com/pod-product-compliance
Lightning Source LLC
Chambersburg PA
CBHW042311210326
41598CB00041B/7355